Sascha Schnepp

Space-Time Adaptive Numerical Methods

Sascha Schnepp

Space-Time Adaptive Numerical Methods

Applied to the Simulation of Charged Particle Beams

Südwestdeutscher Verlag für Hochschulschriften

Impressum/Imprint (nur für Deutschland/ only for Germany)
Bibliografische Information der Deutschen Nationalbibliothek: Die Deutsche Nationalbibliothek verzeichnet diese Publikation in der Deutschen Nationalbibliografie; detaillierte bibliografische Daten sind im Internet über http://dnb.d-nb.de abrufbar.
Alle in diesem Buch genannten Marken und Produktnamen unterliegen warenzeichen-, marken- oder patentrechtlichem Schutz bzw. sind Warenzeichen oder eingetragene Warenzeichen der jeweiligen Inhaber. Die Wiedergabe von Marken, Produktnamen, Gebrauchsnamen, Handelsnamen, Warenbezeichnungen u.s.w. in diesem Werk berechtigt auch ohne besondere Kennzeichnung nicht zu der Annahme, dass solche Namen im Sinne der Warenzeichen- und Markenschutzgesetzgebung als frei zu betrachten wären und daher von jedermann benutzt werden dürften.

Verlag: Südwestdeutscher Verlag für Hochschulschriften Aktiengesellschaft & Co. KG
Dudweiler Landstr. 99, 66123 Saarbrücken, Deutschland
Telefon +49 681 37 20 271-1, Telefax +49 681 37 20 271-0, Email: info@svh-verlag.de
Zugl.: Darmstadt, TU, Diss., 2009

Herstellung in Deutschland:
Schaltungsdienst Lange o.H.G., Berlin
Books on Demand GmbH, Norderstedt
Reha GmbH, Saarbrücken
Amazon Distribution GmbH, Leipzig
ISBN: 978-3-8381-0666-3

Imprint (only for USA, GB)
Bibliographic information published by the Deutsche Nationalbibliothek: The Deutsche Nationalbibliothek lists this publication in the Deutsche Nationalbibliografie; detailed bibliographic data are available in the Internet at http://dnb.d-nb.de.
Any brand names and product names mentioned in this book are subject to trademark, brand or patent protection and are trademarks or registered trademarks of their respective holders. The use of brand names, product names, common names, trade names, product descriptions etc. even without a particular marking in this works is in no way to be construed to mean that such names may be regarded as unrestricted in respect of trademark and brand protection legislation and could thus be used by anyone.

Publisher:
Südwestdeutscher Verlag für Hochschulschriften Aktiengesellschaft & Co. KG
Dudweiler Landstr. 99, 66123 Saarbrücken, Germany
Phone +49 681 37 20 271-1, Fax +49 681 37 20 271-0, Email: info@svh-verlag.de

Copyright © 2009 by the author and Südwestdeutscher Verlag für Hochschulschriften Aktiengesellschaft & Co. KG and licensors
All rights reserved. Saarbrücken 2009

Printed in the U.S.A.
Printed in the U.K. by (see last page)
ISBN: 978-3-8381-0666-3

CONTENTS

Abstract 1

1. Introduction 3
   1.1 Motivation and Objectives 3
   1.2 Outline 5

2. Continuous Electrodynamics 9
   2.1 MAXWELL's Equations in Continuum 9
       2.1.1 MAXWELL's Equations in Integral and Differential Form 9
       2.1.2 Conservation of Charge 10
       2.1.3 Constitutive Equations 11
       2.1.4 Conservation of Electromagnetic Energy 12
       2.1.5 Boundary Conditions 12
       2.1.6 Wave Equation 13
       2.1.7 Relativistic Transformation 14
   2.2 Charged Particle Dynamics 15
       2.2.1 NEWTON-LORENTZ Equations of Motion 15
       2.2.2 HAMILTONIAN Mechanics 15
       2.2.3 Phase Space 16
       2.2.4 LIOUVILLE's Theorem 17
       2.2.5 Statistical Beam Characterization 17

3. Discrete Electrodynamics 23
   3.1 Discretization of Continuous Space 23
   3.2 Finite Integration Technique 29
       3.2.1 Semidiscrete Formulation 29
       3.2.2 Properties of the Finite Integration Technique 33

3.3		Finite Integration-Finite Volume Hybrid Method	37
	3.3.1	Semidiscrete Finite Volume Formulation in 1D	38
	3.3.2	Properties of Finite Volume Methods in 1D	42
	3.3.3	The Hybrid Scheme in 3D	43
3.4		Discontinuous GALERKIN Method	47
	3.4.1	Semidiscrete Formulation	47
	3.4.2	Properties of the Discontinuous GALERKIN Method	49
3.5		Discretization of Time	58
	3.5.1	Time Integration of HAMILTONIAN Systems	58
	3.5.2	Time Integration of Non-HAMILTONIAN Systems	61
	3.5.3	Time Integration Applying Split Operator Techniques	61
3.6		Consistency, Stability and Convergence	66
3.7		Discrete Charged Particle Dynamics	78
	3.7.1	Approaches to Beam Dynamics Simulations	78
	3.7.2	Particle-In-Cell Method	79

4. Time-Adaptive Grid Refinement Techniques ... 87

4.1		Overview of Existing Works	87
4.2		Time-Adaptive Grid Refinement for the Finite Integration Technique	90
	4.2.1	Interpolation of Grid Voltages	90
	4.2.2	Stability	95
	4.2.3	Charged Particle Dynamics on Adaptive Grids	97
4.3		Time-Adaptive Grid Refinement for the FI-FV Method	104
4.4		Time-Adaptive Grid Refinement for the Discontinuous GALERKIN Method	105
	4.4.1	Projection Based h-Adaptation	105
	4.4.2	Projection Based p-Adaptation	109
	4.4.3	Properties of the Projection Based Adaptations	109
	4.4.4	Automated hp-Adaptivity	113

5. Tests and Applications ... 121

5.1		Sanity Check	121
	5.1.1	Grid Adaptation	121
	5.1.2	Interpolation Techniques	122

	5.1.3	Dispersive Behavior of the FIT, LT-FIT and FI-FV scheme 123
	5.1.4	Dispersive Behavior of the DGM 124
	5.1.5	Adaptive FIT Algorithm 124
	5.1.6	Performance 125
	5.1.7	Automated hp-Adaptive DG Scheme 125
5.2	Comparison and Benchmarking of the Numerical Methods 133
5.3	Self-Consistent Simulation of the PITZ RF Gun 138

6. Summary and Outlook 143

Appendix 147

A. A Charge Conserving PIC-Scheme for 3D Beam Dynamics Simulations . . 149

B. Projection Matrices for DG-h-Adaptation 153

C. Efficient Dynamic Memory Management 155

D. Exemplary *tam*BCI Input File 159

Abbreviations and Symbols 165

List of Figures 173

Bibliography 177

ABSTRACT

This work establishes techniques for adjusting the local spatial resolution of selected numerical methods in a time-adaptive manner. Such techniques are developed within the framework of the Finite Integration Technique (FIT), a hybrid Finite Integration-Finite Volume (FI-FV) Scheme and the Discontinuous Galerkin Method (DGM). While the FIT and the DGM are established methods for the numerical solution of electromagnetic field problems, the FI-FV Scheme has been developed in the context of this work.

The semi-discrete, i.e., discrete in space and continuous in time, as well as the fully discretized formulations of all considered methods are presented. For both formulations of each method, an analysis of the dispersive and dissipative behavior on fixed computational grids is carried out. As a result, asymptotic orders of the dispersion and dissipation errors are established.

Techniques for the determination and modification of the discrete electromagnetic field quantities in locally refined regions are presented for each of the numerical methods. For the FIT and the FI-FV Scheme, adaptations based on linear and third order spline interpolations are presented. The adaptation techniques for the DGM are based on projection operators, which are shown to minimize the adaptation error. The numerical stability of the developed adaptive methods is proven.

The developed algorithms are applied to the self-consistent simulation of charged particle dynamics and electrodynamics. The results of the first design study simulating the complete first section of the Free-Electron Laser in Hamburg (FLASH), taking space charge and structure interactions into account, are presented.

1. INTRODUCTION

The computer simulation of electromagnetic fields and, indeed, any other physical phenomena, puts scientists into a dilemma. On the one hand, the steadily increasing computing power of modern computer hardware allows for performing such detailed simulations that the realistic scenario is closely resembled. On the other hand, the sheer size of the computer models in terms of degrees of freedom (DoF) inevitably reveals the weak points of the employed algorithms. For long-term beam dynamics simulations, a key factor, which severely limits the attainable accuracy is the artificial, i.e. non-physical, dispersive behavior of numerical methods. This thesis is concerned with the development of techniques for the adaptation of the local number of DoF in dedicated regions in order to significantly reduce numerical errors at minimal extra computational costs.

1.1 Motivation and Objectives

The methodology in developing and designing particle accelerators is nowadays dominated by numerical simulations. Every component of a new machine has been designed and optimized in computer simulations before an actual manufacturing process starts. In order to further push the frontiers in particle physics, new experiments requiring higher particle energies are necessary. Alongside with the increase of particle energies goes an increase of the costs for realizing large international projects such as the *International Linear Collider* (ILC) [23], the *X-Ray Free-Electron Laser* (XFEL) [6], or the *Large Hadron Collider* (LHC) [24][1]. The estimated total costs for the XFEL project are 850 million Euros, while the required budgets for the ILC and the LHC reach up to six billion Euros. These numbers accentuate the demand for highly accurate simulation results, which ensure a successful and economic operation of the facilities.

Computer simulations do not only facilitate the design and testing process during the development phase and do, thus, lead to shorter development cycles, but they can also provide information about effects that are not accessible to measurements. The injector section of the *Free-Electron Laser in Hamburg* (FLASH) [1][2] serves as an example. A CAD representation of the injector is shown in the figure 1.1. A very short laser pulse releases electrons from the photo cathode. The resonant cavity is excited with a radio frequency electromagnetic field, which accelerates the electrons. They drift through the coupling antenna and the following shutter valve. It is not until the electrons reach the diagnostics section that measurement equipment can be used for characterizing the bunch. The measurements only provide information about

---

[1] To the current date, only the construction of the LHC is finished. It is currently in the start-up phase. While the XFEL has entered the construction phase, it is uncertain whether the ILC will ever be built.
[2] FLASH is a free-electron laser installed at the *Deutsche Elektronen-Synchrotron* (DESY). It can be considered to be the predecessor of the XFEL which will use the same design for the injector section.

the state of the bunch at this position as a result of all influences it experienced before. They do not allow for specifying the dedicated effects of the injector elements. This is only possible by means of numerical simulations. Such simulations have been accomplished and their results are presented in section 5.3.

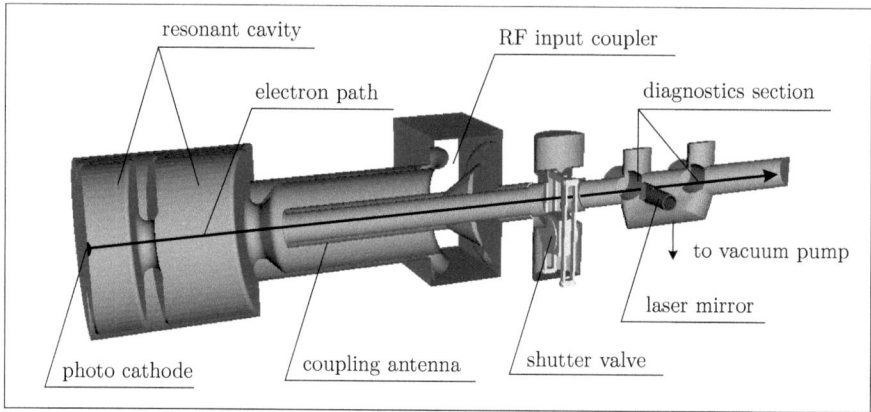

Figure 1.1: FLASH injector section: The figure shows a CAD model of the injector section of the *Free-electron LASer in Hamburg* (FLASH) in cut view. A short laser pulse is directed onto the photo cathode, where electrons are emitted due to the photo effect. The bunch of electrons is accelerated by a high-frequency electromagnetic field, which is excited in the resonant cavity. It propagates along the electron path through the coupling antenna and passes the shutter valve and the laser mirror, which is inserted through one arm of the diagnostics section. The other arms are used for inserting measurement devices. The opening on its bottom side is connected to a vacuum pump.

A variety of computer programs for performing beam dynamics simulations is available[3]. These simulations require the simultaneous solution of the MAXWELL's equations, which cover the macroscopic effects of electromagnetism, and the NEWTON-LORENTZ equations for describing the particle mechanics. However, the solution of this coupled problem can be very expensive in terms of computing time and computer memory, and sometimes it is simply not feasible. It is therefore that simplifications have to be introduced. In some programs the accelerator structure is omitted and the evolution of the particles is considered in free space instead. From the illustration of the complex structure of the FLASH injector, with all its obstacles close to the electron path, it is evident that accurate and realistic results cannot be guaranteed in this case. Other simplifications assume the electromagnetic fields to be static and determined only by the distribution of the charges in space. Since this neglects all time-dependent effects the results are inherently erroneous.

In order to obtain accurate and realistic results, *self-consistent* simulations, which take the surrounding structure and the full set of the coupled MAXWELL and NEWTON-LORENTZ equations into account, have to be performed. Numerical methods such as the *Finite Difference*

---

[3] A short survey of existing programs and approaches is given in section 3.7.

Method (FDM) [124], the *Finite Integration Technique* (FIT) [136], the *Finite Volume Method* (FVM) [83], and *Finite Element Methods* (FEM) [21, 65, 150] are viable choices for fulfilling this task. They make use of a dissection of the continuous space into non-overlapping cells which define a computational grid. The assignment of the characteristic material parameters, permittivity, permeability and conductivity, to all cells leads to a discrete representation of the structure. The continuous electromagnetic quantities are discretized, yielding a finite number of numerical quantities. The discretization errors, that are inherent to those methods, decrease as the number of grid cells increases. However, the available computer memory and the time necessary for solving the problem set an upper limit to the applicable number of cells.

In the figure 1.2 a cut of the FLASH injector is shown in frontal view. The illustration includes the electron bunch, which is drawn approximately on scale. The length of the bunch is about 5 mm while the model has a total length of 2 m. Hence, the geometrical dimensions of the injector and the bunch differ by almost three orders of magnitude. Since the maximum size of the grid cells is prescribed by the smallest detail that has to be resolved, very large numbers of grid cells have to be employed for the simulation of such multi-scale problems. For the given case of a moving, highly localized particle bunch, however, a high grid resolution is required only in the vicinity of the bunch. This is illustrated in the enlargement in figure 1.2. An adaptation of the grid in time would drastically increase the efficiency of the computations. However, in the standard formulations of the numerical methods named above, the grid is not allowed to vary with time. The objective of this thesis is to extend these methods with capabilities for dealing with time-adaptive grid refinement in order to perform efficient self-consistent beam dynamics simulations.

For the Finite Volume Method and the Finite Element Method, time-adaptive grid refinement techniques have been published in, e.g., [13, 28, 45]. In the framework of the Finite Integration Technique, however, it is a new development. The application of time-adaptive grid refinement techniques to transient electromagnetic problems with self-consistent beam dynamics simulations has not been reported for any of the methods.

1.2 Outline

In chapter 2 the governing equations in continuous space and time are established. Section 2.1 is concerned with MAXWELL's equations, which are axiomatically introduced. The following sections focus on properties, such as the conservation of charge and energy, that play an important role in the subsequent derivation and analysis of methods for their numerical solution. The dynamics of charged particles are the subject of section 2.2. Their motion in the presence of electromagnetic fields is covered by the NEWTON-LORENTZ equations. The section finishes with the definition of a measure for the quality of a particle beam and conditions under which this measure is conserved.

Chapter 3 deals with methods for the numerical solution of the combined system of electromagnetic fields and charged particles. Following an introduction of terms and properties relating to computational grids and types of grid refinements, the spatial discretization procedures according to the FIT, a new hybrid FI-FV scheme, and the Discontinuous GALERKIN FEM (DGM) are addressed in the sections 3.1–3.4. For each of these methods, an analysis comprising their conservation and dispersion properties is carried out. The key terms for describing numerical

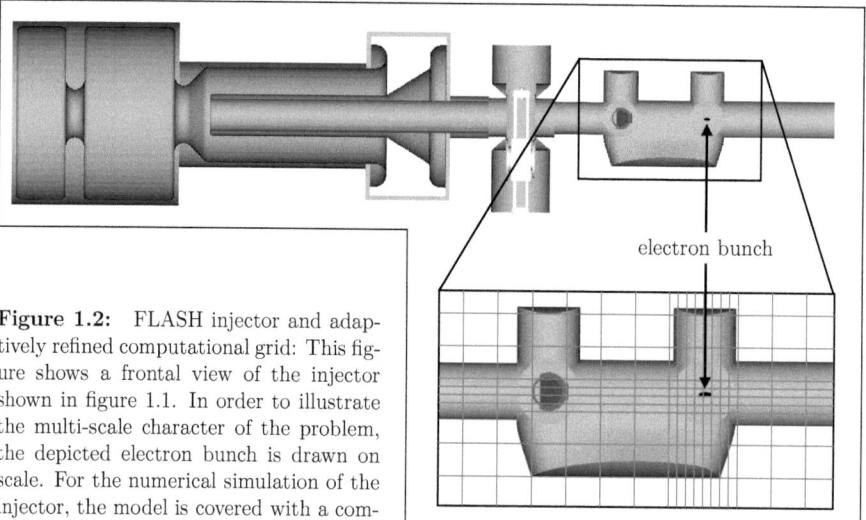

Figure 1.2: FLASH injector and adaptively refined computational grid: This figure shows a frontal view of the injector shown in figure 1.1. In order to illustrate the multi-scale character of the problem, the depicted electron bunch is drawn on scale. For the numerical simulation of the injector, the model is covered with a computational grid. All physical quantities are cell-wise discretized. This yields a finite number of numerical quantities which can be processed in computer simulations. Since the size of the grid cells is determined by the objects that have to be resolved, the application of an adaptive grid refinement which tracks the particle bunch is an emerging idea. The development of techniques for time-adaptive grid refinement and its application in beam dynamics simulations is the objective of this thesis.

methods, *consistency, stability* and *convergence*, are defined in section 3.6. Asymptotic convergence orders of the spatially discrete equations of the FIT, the FI-FV scheme, and the DGM regarding their dispersive and dissipative properties along a coordinate are established. In addition, an analysis of the accuracy of the fully discretized equations is carried out. Section 3.7 is devoted to the self-consistent incorporation of charged particle dynamics.

In chapter 4, techniques for the adaptation of the local resolution for all presented methods are established. Within the FIT and the FI-FV scheme this adaptation is realized via increasing or decreasing the local grid line density. In addition to manipulating the computational grid, the DGM offers the possibility of enriching or reducing the space of functions which is employed for the field approximation. The detailed description of the adaptation algorithms is followed by proofs of the stability for all adaptation techniques. The chapter concludes with the presentation of a method for performing automatic resolution adaptations with the DGM.

Throughout the chapters 3 and 4 the focus is on developing methods for the efficient and accurate long-term simulation of self-consistent beam dynamics. Nevertheless, the methods are presented in a sufficiently general way, such that they can easily be adapted to other application areas.

The developed algorithms are tested, benchmarked and then applied to the simulation of the FLASH injector. This is presented in chapter 5. First, a sanity check of individual parts of

1.2. Outline

the algorithms is performed, in order to ensure their correct functionality. The propagation of a particle bunch in a metallic tube is employed as a benchmark example. All methods are applied to this example using various settings. The results are compared to its analytical solution. Finally, the results of a design study of the FLASH injector are presented.

Chapter 6 summarizes the most important achievements and results of this thesis and depicts some topics for further research.

2. CONTINUOUS ELECTRODYNAMICS

The set of MAXWELL's equations describes all macroscopic phenomena of electromagnetism. Their integral and differential forms are axiomatically stated in the first section of this chapter. With these equations at hand, conservation properties, boundary conditions and the wave equation are addressed.

The second section of this chapter is concerned with charged particle dynamics. Charged particles are, at the same time, subject to the electromagnetic field forces as well as sources of such fields. The particle motion is governed by the NEWTON-LORENTZ equations, which couple to MAXWELL's equations via the LORENTZ force. The phase space is introduced as a convenient system for describing the particle dynamics. It allows for the characterization of particle beams by means of statistical variables. The section finishes with a discussion on the quality of particle beams and on conditions for its preservation.

2.1 Maxwell's Equations in Continuum

Starting from 1861 James Clerk MAXWELL published a set of four equations which cover the macroscopic phenomena of electromagnetism [85–87]. They combine the previously existing works of AMPÈRE [7] and FARADAY [51] as well as experimental results known at the time. The introduction of the dielectric displacement completed MAXWELL's theory of electromagnetism which is valid up to the present day.

2.1.1 Maxwell's Equations in Integral and Differential Form

MAXWELL's equations in integral form read:

$$\int_{\partial A} \vec{H}(\vec{r},t) \cdot \mathrm{d}\vec{s} = \int_{A} \left(\partial_t \vec{D}(\vec{r},t) + \vec{J}(\vec{r},t) \right) \cdot \mathrm{d}\vec{A} \qquad (2.1)$$

$$\int_{\partial A} \vec{E}(\vec{r},t) \cdot \mathrm{d}\vec{s} = -\int_{A} \partial_t \vec{B}(\vec{r},t) \cdot \mathrm{d}\vec{A} \qquad (2.2)$$

$$\int_{\partial V} \vec{D}(\vec{r},t) \cdot \mathrm{d}\vec{A} = \int_{V} \rho(\vec{r},t) \, \mathrm{d}V \qquad (2.3)$$

$$\int_{\partial V} \vec{B}(\vec{r},t) \cdot \mathrm{d}\vec{A} = 0 \qquad (2.4)$$

for any surface A and volume V. The spatial variable is denoted by $\vec{r}$, and the temporal variable by t. The magnetic field strength is indicated by $\vec{H}$, $\vec{B}$ denotes the magnetic flux density, $\vec{E}$ the electric field strength, $\vec{D}$ the dielectric flux density or dielectric displacement, and $\vec{J}$ the electric current density. The current density can be decomposed into a conductive part $\vec{J}_\kappa$, a convective part $\vec{J}_c$ and the impressed current density $\vec{J}_e$.

Throughout this work, the *SI system of units* is used. The units of the electromagnetic field quantities are:

$$\begin{aligned}
[E] &= \text{V/m} &&= \text{Volt / meter} \\
[D] &= \text{As/m}^2 &&= \text{Ampère second / meter}^2 \\
[H] &= \text{A/m} &&= \text{Ampère / meter} \\
[B] &= \text{Vs/m}^2 &&= \text{Volt second / meter}^2 \\
[J] &= \text{A/m}^2 &&= \text{Ampère / meter}^2
\end{aligned} \quad (2.5)$$

Typically, the unit of the magnetic flux density $\vec{B}$ is named Tesla ($1\,\text{T} = 1\,\text{Vs/m}^2$).

Equation (2.1) is referred to as AMPÈRE's *law*, equation (2.2) is referred to as FARADAY's *law* and the equations (2.3) and (2.4) are named GAUSS' *law* and GAUSS' *law of magnetism*. The latter states the absence of magnetic charges.

For stationary media, MAXWELL's equations can be expressed equivalently in integral and differential form using the general STOKES theorem [20]. Applying the KELVIN-STOKES theorem to AMPÈRE's and FARADAY's law yields their differential forms:

$$\nabla \times \vec{H} = \partial_t \vec{D} + \vec{J} \quad (2.6)$$
$$\nabla \times \vec{E} = -\partial_t \vec{B} \quad (2.7)$$

Above, the spatial and temporal dependences of the electromagnetic field quantities are implicitly assumed. In figure 2.1 the relation of the flux and field quantities given by the two equations is illustrated.

Applying GAUSS-OSTROGRADSKY's theorem to the equations (2.3) and (2.4) yields their differential forms:

$$\nabla \cdot \vec{D} = \rho \quad (2.8)$$
$$\nabla \cdot \vec{B} = 0 \quad (2.9)$$

This thesis deals with transient problems involving freely moving charges. It does, therefore, focus on the time-dependent MAXWELL's equations.

2.1.2 Conservation of Charge

In order to demonstrate the conservation of charge, the divergence operator is applied to AMPÈRE's and FARADAY's law leading to:

$$\nabla \cdot (\nabla \times \vec{H}) = \partial_t \nabla \cdot \vec{D} + \nabla \cdot \vec{J} \quad (2.10)$$
$$\nabla \cdot (\nabla \times \vec{E}) = -\partial_t \nabla \cdot \vec{B} \quad (2.11)$$

Using the vector calculus identity $\nabla \cdot (\nabla \times \vec{a}) \equiv 0$, one obtains:

$$0 = \partial_t \nabla \cdot \vec{D} + \nabla \cdot \vec{J} \quad (2.12)$$
$$0 = \partial_t \nabla \cdot \vec{B} \quad (2.13)$$

2.1. MAXWELL's Equations in Continuum

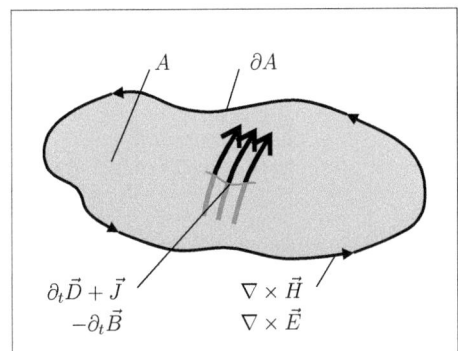

Figure 2.1: Illustration of AMPÈRE's and FARADAY's law: The equations (2.1), (2.2) and likewise (2.6), (2.7) relate the temporal change of the flux quantities penetrating the surface A to the field quantities on the boundary of the surface ∂A.

Inserting GAUSS' law into (2.12) reveals the charge conservation law:

$$\partial_t \rho + \nabla \cdot \vec{J} = 0 \qquad (2.14)$$

Equation (2.14) is also called the *continuity equation* of electric charge and current density. It reflects the physical reality that a temporal change of the total charge Q in a volume V is caused only by a flow of electric current $\vec{I}$ into, or out of the volume V.

2.1.3 Constitutive Equations

The fields and fluxes appearing in (2.1)-(2.4) are connected by constitutive equations. In this thesis linear, isotropic, non-dispersive and time-independent materials are assumed. For this case the constitutive equations read:

$$\vec{B}(\vec{r},t) = \mu(\vec{r})\,\vec{H}(\vec{r},t) \qquad (2.15)$$
$$\vec{D}(\vec{r},t) = \epsilon(\vec{r})\,\vec{E}(\vec{r},t) \qquad (2.16)$$
$$\vec{J}_\kappa(\vec{r},t) = \kappa(\vec{r})\,\vec{E}(\vec{r},t) \qquad (2.17)$$

Equation (2.17) is referred to as OHM's law. These additional equations make MAXWELL's equations solvable.

The parameters μ, ϵ and κ are the electromagnetic characteristics of materials. They are called *permeability, permittivity* and *conductivity*. The permeability and permittivity of a material are commonly given relative to the respective values in vacuum μ_0 and ϵ_0:

$$\mu(\vec{r}) = \mu_0\,\mu_r(\vec{r}) \quad \text{with} \quad \mu_0 = 4\pi \cdot 10^{-7}\ \text{Vs/Am} \qquad (2.18)$$
$$\epsilon(\vec{r}) = \epsilon_0\,\epsilon_r(\vec{r}) \quad \text{with} \quad \epsilon_0 = \frac{1}{\mu_0 c_0^2} \approx 8.854 \cdot 10^{-12}\ \text{As/Vm} \qquad (2.19)$$

where $c_0 \approx 2.998 \cdot 10^8$ m/s is the speed of light in vacuum.

2.1.4 Conservation of Electromagnetic Energy

For linear, non-dispersive materials as assumed in the preceding section, the electromagnetic *energy density* w is given by:

$$w(\vec{r},t) = \frac{1}{2}\left(\vec{E}(\vec{r},t)\cdot\vec{D}(\vec{r},t) + \vec{B}(\vec{r},t)\cdot\vec{H}(\vec{r},t)\right) \tag{2.20}$$

Integrating the energy density over a domain Ω yields the total electromagnetic energy W as a function of time:

$$W(t) = \int_\Omega w(\vec{r},t)\,\mathrm{d}V \tag{2.21}$$

The conservation of the total electromagnetic energy, can be derived from POYNTING's *theorem*:

$$\partial_t w + \nabla\cdot\vec{S} = -\vec{J}\cdot\vec{E} \tag{2.22}$$

The POYNTING vector $\vec{S}$ represents an energy flux ([71] p. 259) and is given by:

$$\vec{S} := \vec{E}\times\vec{H} \tag{2.23}$$

Assuming the absence of currents, the term $\vec{J}\cdot\vec{E}$ in (2.22) vanishes. If, moreover, the domain Ω is closed and bounded in such a way that fields cannot penetrate the boundary, it follows:

$$\mathrm{d}_t\int_\Omega w(\vec{r},t)\,\mathrm{d}V = \mathrm{d}_t W(t) = -\int_\Omega \nabla\cdot\vec{S}\,\mathrm{d}V = 0 \tag{2.24}$$

In that case, the local energy density w may vary with time but the total electromagnetic energy remains constant:

$$W(t) = W \tag{2.25}$$

Taking currents in the domain into account, the term $\vec{J}\cdot\vec{E}$ corresponds to a power density. If we assume the currents to be purely convective $\vec{J} = \vec{J}_c$, i.e., arising from the motion of free charges, this implies a transfer of electromagnetic field energy to kinetic energy of the charged particles and vice versa. The responsible force is the so-called LORENTZ *force*:

$$\vec{F} = q\left(\vec{E} + \vec{v}\times\vec{B}\right) \tag{2.26}$$

where q is the charge of the particle. Since the force exerted by the magnetic field is oriented perpendicularly to the velocity vector $\vec{v}$, it does not change the particles' energy. Therefore, only the electric field appears on the right hand side of POYNTING's theorem.

2.1.5 Boundary Conditions

The electric and magnetic field quantities obey boundary conditions at interfaces between two media, 1 and 2, with different material characteristics. The normal component of the electric and magnetic flux density, $\vec{D}$ and $\vec{B}$, on either side of the material interface are related by:

$$(\vec{D}_2 - \vec{D}_1)\cdot\vec{n} = \sigma_\mathrm{I} \tag{2.27}$$
$$(\vec{B}_2 - \vec{B}_1)\cdot\vec{n} = 0 \tag{2.28}$$

2.1. Maxwell's Equations in Continuum

with the unit normal vector $\vec{n}$ of the interface pointing from medium 1 to medium 2 and the idealized interface charge density σ_I.

For the tangential component of the electric and magnetic field strength, $\vec{E}$ and $\vec{H}$, the relation is given by:

$$\vec{n} \times (\vec{E}_2 - \vec{E}_1) = 0 \tag{2.29}$$
$$\vec{n} \times (\vec{H}_2 - \vec{H}_1) = J_\mathrm{I} \tag{2.30}$$

with the idealized interface current density J_I. The normal magnetic flux density as well as the tangential electric field strength are, therefore, continuous at material interfaces. Assuming a perfectly electrically conducting (PEC) material, i.e., a material with a conductivity of $\kappa \to \infty$, this material has to be free of any electric field due to equation (2.17). In any other case a non-physical infinite current density would occur. Moreover, due to Faraday's law, no time-varying magnetic field can exist in such a material.

Complemented by the boundary conditions above and suitable initial conditions, a solution to Maxwell's equations exists, and it is unique [71, 82].

2.1.6 Wave Equation

Ampère's as well as Faraday's equation describe transient phenomena, i.e., phenomena that show a time dependent behavior. Applying the curl operator to the equations (2.6), (2.7) and using the vector calculus relation:

$$\nabla \times (\nabla \times \vec{a}) = \nabla(\nabla \cdot \vec{a}) - \Delta \vec{a} \tag{2.31}$$

the homogeneous, second order electric or magnetic wave equation (without source terms) is derived:

$$\Delta U - \frac{1}{c^2} \partial_t^2 U = 0 \quad \text{with} \quad U \in \{\vec{E}, \vec{H}\} \tag{2.32}$$

with the local speed of light:

$$c = (\epsilon \mu)^{-1/2} \tag{2.33}$$

This reveals the hyperbolic character of the equations (2.1), (2.2) and (2.6), (2.7) respectively.

Applying the Fourier transform in space, $\mathscr{F}_r$, to the wave equation yields the ordinary differential equation (ODE) ([38] pp. 25):

$$\frac{\mathrm{d}^2 \hat{u}(\vec{k}, t)}{\mathrm{d}t^2} + c^2 |\vec{k}|^2 \hat{u}(\vec{k}, t) = 0 \tag{2.34}$$

with:

$$\mathscr{F}_r(U(\vec{r}, t)) = \hat{u}(\vec{k}, t) \tag{2.35}$$

Above, $\vec{k}$ denotes the *wave vector*. The solutions of this ODE are of the form:

$$\hat{u}(\vec{k}, t) = A(\vec{k}) e^{\mathrm{i} c |\vec{k}| t} + B(\vec{k}) e^{-\mathrm{i} c |\vec{k}| t} \tag{2.36}$$

where A and B are real-valued functions of $\vec{k}$. Applying the Fourier transform in time, $\mathscr{F}_t$, to the ODE (2.34) with:

$$\mathscr{F}_t\left(\hat{u}(\vec{k}, t)\right) = \hat{u}(\vec{k}, \omega) \tag{2.37}$$

and ω the *angular frequency*, yields the *dispersion relation* of the homogeneous wave equation (2.32):
$$\omega^2 = c^2|\vec{k}|^2 = c^2\left(k_x^2 + k_y^2 + k_z^2\right) \tag{2.38}$$
The solutions (2.36) can be rewritten as:
$$\hat{u}(\vec{k},t) = A(\vec{k})\mathrm{e}^{\mathrm{i}\omega t} + B(\vec{k})\mathrm{e}^{-\mathrm{i}\omega t} \tag{2.39}$$
Performing the inverse FOURIER transform in space finally provides solutions in the form:
$$U(\vec{r},t) = \frac{1}{(2\pi)^{3/2}} \int_{\mathbb{R}^3} \left[A(\vec{k})\mathrm{e}^{\mathrm{i}(\omega t + \vec{k}\cdot\vec{r})} + B(\vec{k})\mathrm{e}^{\mathrm{i}(-\omega t + \vec{k}\cdot\vec{r})} \right] \mathrm{d}\vec{k} \tag{2.40}$$
Hence, the solutions of the homogeneous wave equation (2.32) can be expressed by a superposition of plane waves.

2.1.7 Relativistic Transformation

MAXWELL's equations are defined on a four-dimensional space-time domain. They were introduced using the spatial coordinate $\vec{r}$ from the three-dimensional spatial domain Ω and the temporal coordinate t defined on the one-dimensional temporal domain T. The transformation of the coordinates of a point between two coordinate systems, which have a relative velocity is described by the LORENTZ transformation.

The coordinates in the two systems are given by (ct, x, y, z) and (ct', x', y', z'). It is assumed that at the time $t = t' = 0$ the frames share a common origin and the coordinate axes are aligned. For a relative velocity of $\vec{v} = (0, 0, v_z)$ the transformation reads:
$$\begin{pmatrix} ct' \\ x' \\ y' \\ z' \end{pmatrix} = \begin{pmatrix} \gamma & 0 & 0 & -\beta\gamma \\ 0 & 1 & 0 & 0 \\ 0 & 0 & 1 & 0 \\ -\beta\gamma & 0 & 0 & \gamma \end{pmatrix} \begin{pmatrix} ct \\ x \\ y \\ z \end{pmatrix} \tag{2.41}$$
with the relative velocity factor β and the LORENTZ *factor* γ defined in the following way:
$$\beta := \frac{|\vec{v}|}{c} \tag{2.42}$$
$$\gamma := \frac{1}{\sqrt{1-\beta^2}} \tag{2.43}$$
The LORENTZ factor γ is also called the *relativistic contraction factor*. Equation (2.41) shows a coupling of the time coordinate ct to the respective spatial coordinate of the direction of relative motion z and vice versa. The spatial coordinates oriented perpendicularly to the direction of relative motion, x and y, do not experience a transformation. LORENTZ' transformation for arbitrary relative motion is given in [71] pp. 539.

The consequence of this transformation is reflected, e.g., in the contraction of the electromagnetic field of a point charge, traveling at a velocity $\vec{v}$. At rest, the purely electric field of a point charge shows a spherical symmetry. If, however, the point charge moves with a velocity $v \to c$ its fields, observed from a point at rest, are contracted to an infinitely flat plane. The normal vector $\vec{n}$ of the plane and the velocity vector $\vec{v}$ are aligned. Moreover, the field is not purely electric anymore, but it has an azimuthal magnetic component as well. A detailed derivation is found in [62] and [103].

2.2 Charged Particle Dynamics

In this section the motion of charged particles in the presence of electromagnetic fields is considered. The change of a particle's momentum and energy is due to the LORENTZ force and obeys NEWTON's second law of classical mechanics. After introducing the corresponding forces in section 2.2.1, the formulation of the problem as a HAMILTONIAN system is presented in section 2.2.2. A convenient description of the state of a mechanical system, arising from the HAMILTONIAN, is the phase space which is introduced in section 2.2.3. LIOUVILLE's theorem formulates a key limitation in the design of particle accelerators. The theorem is stated in section 2.2.4. Also in this section, the characterization of charged particle beams by means of moments of the particle distribution in phase space is presented. Finally, in section 2.2.5, a measure for the quality of a particle beam is introduced, and the significance of LIOUVILLE's theorem for preserving this quality is explained.

2.2.1 Newton-Lorentz Equations of Motion

The link between MAXWELL's equations and the equations of motion of a charged body is established by the LORENTZ force, defined in equation (2.26), on the one hand, and NEWTON's second law:

$$\vec{F} = \mathrm{d}_t \vec{p} = \mathrm{d}_t(m\vec{v}) \qquad (2.44)$$

with the mechanical momentum $\vec{p}$, on the other hand. In the relativistic energy regime, the mass of the particle m is related to the contraction factor γ by:

$$m(\gamma) = \gamma m_0 \qquad (2.45)$$

with m_0 the mass of the particle at rest. Inserting the LORENTZ force into NEWTON's second law yields the rate of change of particle momentum as a function of the electric and the magnetic field:

$$\mathrm{d}_t \vec{p} = q\left(\vec{E} + \vec{v} \times \vec{B}\right) \qquad (2.46)$$

2.2.2 Hamiltonian Mechanics

HAMILTONIAN mechanics is a reformulation of classical mechanics, utilizing *generalized coordinates*[1] and *momenta* and a set of first order differential equations. The generalized coordinates are denoted by $\mathcal{Q}_i(t)$ and the generalized canonical momenta are denoted by $\mathcal{P}_i(t)$ for $i = 1..I$ with I the number of degrees of freedom. The HAMILTON equations are given by:

$$\mathrm{d}_t \mathcal{Q}_i(t) = \partial_{\mathcal{P}_i} \mathcal{H}(\mathcal{Q}_i, \mathcal{P}_i, t) \qquad (2.47)$$
$$\mathrm{d}_t \mathcal{P}_i(t) = -\partial_{\mathcal{Q}_i} \mathcal{H}(\mathcal{Q}_i, \mathcal{P}_i, t) \qquad (2.48)$$

where $\mathcal{H}(\mathcal{Q}_i, \mathcal{P}_i, t)$ is the HAMILTONIAN. If $\mathcal{H}$ does not explicitly depend on the temporal variable t, it corresponds to the total energy of the system under consideration.

---

[1] Generalized coordinates are unspecified coordinates which can be replaced by a specific coordinate system after the derivation of the equations of motion.

The canonical momentum $\mathcal{P}$, conjugate to the specific Cartesian coordinate r, is given by (see [71] p. 582):

$$\mathcal{P} = \vec{p} + \frac{q}{c}\vec{A} \tag{2.49}$$

with the kinetic momentum given in equation (2.44) and $\vec{A}$ the magnetic vector potential, characterized by:

$$\vec{B} = \nabla \times \vec{A} \tag{2.50}$$

A thorough description of HAMILTONIAN mechanics is found in the book of LANDAU and LIFSHITZ [76] and in [26, 141] with a focus on charged particle dynamics.

2.2.3 Phase Space

The possible states of a physical system obeying HAMILTON's equations (2.47) and (2.48) define the *phase space* $\mathbb{S}$. It is a $2I$-dimensional space with coordinate axes corresponding to the I generalized coordinates and the I conjugate momenta. In the following, Cartesian coordinates are considered. A single particle in three-dimensional space has six degrees of freedom (DoF) $(x, y, z, \mathcal{P}_x, \mathcal{P}_y, \mathcal{P}_z)$, resulting in the six-dimensional phase space $\mathbb{S}^6$. Consequently, a system composed of N particles has $6N$ DoF. The phase space distribution of the particles can be represented by a *density function* $\varrho^N(u_1, .., u_N, \mathcal{P}_1, .., \mathcal{P}_N)$ with:

$$\int \varrho^N \, du_1 \cdot \ldots \cdot du_N \cdot d\mathcal{P}_1 \cdot \ldots \cdot d\mathcal{P}_N = 1 \tag{2.51}$$

For indistinguishable particles, the phase space distribution can by described by the density function:

$$\varrho^1(u_1, \mathcal{P}_1) = \int \varrho^N \, du_2 \cdot \ldots \cdot du_N \cdot d\mathcal{P}_2 \cdot \ldots \cdot d\mathcal{P}_N \tag{2.52}$$

which is defined in $\mathbb{S}^6$.

In particle accelerators, the longitudinal motion of the particles is, in general, considered to be decoupled from their transverse motion[2]. This allows for a dissection of the six-dimensional phase space $\mathbb{S}^6$ into a set of subspaces of lower dimensionality $\{\mathbb{S}_L^2, \mathbb{S}_T^4\}$. The ultimate reduction of the phase space dimensionality would occur if the motion along any coordinate axes is decoupled from any other direction, resulting in three independent two-dimensional phase spaces. However, the transverse directions of motion are, in general, coupled.

In accelerator physics, the generalized coordinates are commonly chosen such that the longitudinal coordinate coincides with the design orbit. In a circular machine it is, therefore, a curvilinear coordinate, usually denoted by s. The transverse coordinates are named *horizontal* and *vertical*. The horizontal coordinate x is directed parallel to the surface of the ground, the vertical coordinate y perpendicular to it [142]. The examples, which are presented in chapter 5, comprise straight accelerator sections only. The longitudinal coordinate is, therefore, called z-coordinate.

In order to investigate the horizontal or vertical properties of a bunch of particles, a projection of the transverse phase space $\mathbb{S}_T^4$ to the corresponding horizontal or vertical phase space ($\mathbb{S}_H^2, \mathbb{S}_V^2$) is performed. An illustrative phase space diagram is shown in figure 2.2.

---

[2] The longitudinal motion of a particle, is specified as the motion parallel to the so-called *design orbit*. The motion in the plane perpendicular to the design orbit is referred to as transverse motion [81, 102].

2.2.4 Liouville's Theorem

An important property of a distribution in the phase space is described by LIOUVILLE's theorem.

Theorem 2.1 (LIOUVILLE's Theorem). *An infinitesimal element of volume in $2I$-dimensional phase space dV is given by:*

$$dV = d\mathcal{Q}_1 \cdot \ldots \cdot d\mathcal{Q}_I \cdot d\mathcal{P}_1 \cdot \ldots \cdot d\mathcal{P}_I \tag{2.53}$$

Then the following holds true:

$$\int dV = \text{constant} \tag{2.54}$$

Associating a local density, ϱ, in phase space with the volume element dV, this density is conserved. Therefore, the phase space distribution of particle beams cannot be arbitrarily compressed. The impact of this on particle accelerator physics is described in more details in the next section.

LIOUVILLE's theorem does not apply to projected phase spaces, e.g., $\mathbb{S}_H^2, \mathbb{S}_V^2$, defined in the preceding section. The occupied volume in projected phase spaces is not a conserved quantity.

2.2.5 Statistical Beam Characterization

A (bunched) particle beam consists of a large number of particles within a relatively small volume. Ideally, all particles belong to the same species, e.g., electrons. Their phase space distribution can, thus, be described by means of the phase space density function $\varrho^1(u, u')$. The moments of this function are employed for characterizing particle beams and for measuring their quality.

k-m-th Order Moments

In practice, a beam is characterized in the two-dimensional phase spaces $\mathbb{S}_H^2, \mathbb{S}_V^2$ and $\mathbb{S}_L^2$. The bivariate moments of a distribution are, therefore, introduced using the coordinate pair $(u, u') \in \mathbb{S}^2$ with $u \in \{\mathcal{Q}_x, \mathcal{Q}_y, \mathcal{Q}_z\}$ and $u' \in \{\mathcal{P}_x, \mathcal{P}_y, \mathcal{P}_z\}$ [25, 96]. The explicit time dependence of the moments is omitted.

The k-m-th order *raw moment* $\hat{\mu}^{(k,m)}$ of $\varrho^1(u, u')$ is given by:

$$\hat{\mu}^{(k,m)} := E\{(u)^k (u')^m\} = \int_{-\infty}^{\infty} \int_{-\infty}^{\infty} (u)^k (u')^m \varrho^1(u, u') \, du \, du' \tag{2.55}$$

where $E\{\cdot\}$ denotes the expectation value. The linear moments of position and momentum $\hat{\mu}^{(1,0)}$, $\hat{\mu}^{(0,1)}$ are the mean values of the particle distribution. They are simply denoted by μ_u and $\mu_{u'}$.

Taking the moments about the mean yields the *central moments*:

$$\mu^{(k,m)} := E\{(u - \mu_u)^k (u' - \mu_{u'})^m\} = \int_{-\infty}^{\infty} \int_{-\infty}^{\infty} (u - \mu_u)^k (u' - \mu_{u'})^m \varrho^1(u, u') \, du \, du' \tag{2.56}$$

For the second order central moments $\mu^{(2,0)}$ and $\mu^{(0,2)}$ the term *variance* is widely used, with:

$$\mu^{(2,0)} = \sigma_u^2 \tag{2.57}$$
$$\mu^{(0,2)} = \sigma_{u'}^2 \tag{2.58}$$

where σ is the standard deviation. The variance is a measure for the position and momentum spread of a beam. The central moment $\mu^{(1,1)}$ is called *covariance*:

$$\mu^{(1,1)} = \sigma_{u,u'} \tag{2.59}$$

The covariance provides information about the phase space correlation of position and momentum. The variances and the covariance are gathered in the covariance matrix $\mathbf{V}_{u,u'}$, given by:

$$\mathbf{V}_{u,u'} = \begin{pmatrix} \sigma_u^2 & \sigma_{u,u'} \\ \sigma_{u,u'} & \sigma_{u'}^2 \end{pmatrix} \tag{2.60}$$

Beam Quality and Emittance

The quality of a particle beam has to be defined with respect to the particular needs of the driven experiment. Often the figure of merit is the *brightness B*. It relates the beam current to the minimal spot size the beam can be focused to. The higher the brightness is, the higher is the efficiency of an experiment [142]. It is defined by:

$$B := \frac{I}{\pi^2 \varepsilon_x \varepsilon_y} \tag{2.61}$$

with the projected transverse emittances ε_x and ε_y. The *projected emittance* is defined by:

$$\varepsilon_u := \left(\det\left(\mathbf{V}_{u,u'}\right)\right)^{1/2} = \left(\sigma_u^2 \cdot \sigma_{u'}^2 - (\sigma_{u,u'})^2\right)^{1/2} \tag{2.62}$$

This expression is also referred to as the root-mean-square (RMS) emittance [26]. The brightness scales with the inverse of the emittance. Hence, a small emittance identifies a high quality beam.

For every constant ε_u, the equation (2.62) describes a tilted ellipse in the respective projected phase space $\mathbb{S}_u^2$. The occupied area A of the ellipse relates to the emittance by:

$$A = \pi \varepsilon \tag{2.63}$$

In figure 2.2 the phase space ellipse for an exemplary particle distribution in the projected phase space $\mathbb{S}_x^2$ is shown. Figure 2.3 illustrates the connection between the state of a beam and its emittance.

2.2. Charged Particle Dynamics

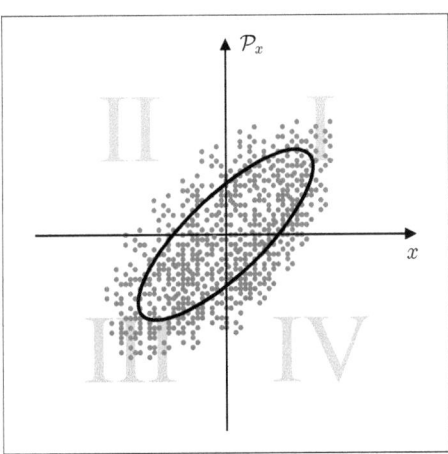

Figure 2.2: Phase space diagram: Portrait of a particle beam in the two-dimensional projected phase space $\mathbb{S}_x^2$. The particle position x is plotted along the abscissa, the particle momentum $\mathcal{P}_x$ along the ordinate. Particles in the quadrants I and III diverge from the beam center, whereas particles in the quadrants II and IV converge to the center. The phase space ellipse is determined by equation (2.62). The area occupied by the ellipse is proportional to the emittance. It is a measure for the quality of a particle beam (see fig. 2.3 for a graphical interpretation).

Several definitions of the emittance are commonly in use. Here, the projected emittance was derived from the canonical phase space. Other sources define the emittance ε^m in an (x, x')-coordinate system using:

$$x' := \frac{p_x}{|\vec{p}|} \tag{2.64}$$

where $\vec{p}$ is the mechanical momentum of the particle. If the emittance is defined in such a way, it experiences an adiabatic damping during beam acceleration [140]. Therefore, the so-called *normalized emittance* ε^n was introduced with:

$$\varepsilon^\mathrm{n} := \beta\gamma\varepsilon^\mathrm{m} \tag{2.65}$$

which is invariant under acceleration. In some definitions the phase space ellipse is not described by the beam variances and its covariance, but it is chosen such that it contains 95 % of the particles in the respective projected phase space[3].

It is common to present values of the emittances in units of $[\varepsilon]$ = mm mrad. Therefore, the RMS emittance is evaluated in (x, x')-coordinates, instead of the canonical coordinates.

In the longitudinal direction, a beam is usually characterized by its RMS length σ_z and the relative RMS *energy spread* $\sigma_{\delta W}$ with $\delta W = (W - \mu_W)/\mu_W$ and W the energy of the particles.

Emittance Conservation

In the preceding section, the brightness B was introduced as a measure of quality for many accelerators. It relates to the inverse of the product of the transverse emittances ε_x and ε_y. The transverse emittances themselves correspond to an area in phase space via equation (2.63). Assuming for the moment that the particle motion in the horizontal plane is decoupled from the motion in the vertical plane, LIOUVILLE's theorem (2.54) states the conservation of these emittances. However, the statement is not generally valid. It is known that, e.g., space charge

---
[3] See [32] p. 50 for a table of factors relating the RMS emittance to other common definitions of the emittance.

forces cause a growth of emittance. Due to the importance of a low emittance for FLASH and the XFEL (see the introduction) as well as for colliders and Free Electron Lasers (FEL) in general, the conditions for emittance conservation are derived in the following.

In the derivation a time-independent HAMILTONIAN $\mathcal{H}$, i.e., a situation with no acceleration, is assumed. Moreover, the particle motion in the two-dimensional phase space $\mathbb{S}_u^2$ is assumed to be decoupled from the motion in the remaining dimensions and, in general, non-linear. For this situation the HAMILTONIAN is given by [26]:

$$\mathcal{H}(u, u') = \frac{(u')^2}{2} + f(u) \qquad (2.66)$$

Substituting the total derivatives with respect to time in HAMILTON's equations ((2.47) and (2.48)) by total derivatives with respect to the longitudinal coordinate s, yields:

$$d_s u = u' \qquad (2.67)$$
$$d_s u' = -\partial_u \mathcal{H}(u, u') \qquad (2.68)$$

The velocity of the frame along the longitudinal coordinate s is w.l.o.g. set to one.

In order for the emittance ε_u to be conserved, its derivative with respect to the longitudinal coordinate s must vanish:

$$d_s \varepsilon_u \stackrel{!}{=} 0 \qquad (2.69)$$

The following relations between raw and central moments are used below:

$$\mu^{(2,0)} = \hat{\mu}^{(2,0)} - (\mu_u)^2 \qquad (2.70)$$
$$\mu^{(0,2)} = \hat{\mu}^{(0,2)} - (\mu_{u'})^2 \qquad (2.71)$$
$$\mu^{(1,1)} = \hat{\mu}^{(1,1)} - \mu_u \mu_{u'} \qquad (2.72)$$

Applying the derivative d_s to the emittance yields:

$$d_s \varepsilon_u = d_s \left(\mu^{(2,0)} \mu^{(0,2)} - \left(\mu^{(1,1)} \right)^2 \right)^{1/2}$$
$$= \frac{1}{2\varepsilon_u} (\underbrace{d_s [\mu^{(2,0)} \mu^{(0,2)}]}_{(I)} - \underbrace{d_s [\left(\mu^{(1,1)} \right)^2]}_{(II)}) \qquad (2.73)$$

Using (2.70) and (2.71), the expression (I) is rewritten as:

$$(I): \quad d_s \left[\mu^{(2,0)} \mu^{(0,2)} \right] = d_s \left[\left(\hat{\mu}^{(2,0)} - (\mu_u)^2 \right) \left(\hat{\mu}^{(0,2)} - (\mu_{u'})^2 \right) \right]$$
$$= \left(d_s \hat{\mu}^{(2,0)} - d_s \left((\mu_u)^2 \right) \right) \mu^{(0,2)} - \mu^{(2,0)} \left(d_s \hat{\mu}(0,2) - d_s \left((\mu_{u'})^2 \right) \right) \qquad (2.74)$$

Exchanging the order of the derivative operator and the expectation value in the definition of the moments (2.55), and inserting HAMILTON's equations yields:

$$(I): \quad d_s \left[\mu^{(2,0)} \mu^{(0,2)} \right] = 2\mu^{(1,1)} \mu^{(0,2)} + 2\mu^{(2,0)} \left(E\{u' \partial_u \mathcal{H}\} - \mu_{u'} E\{\partial_u \mathcal{H}\} \right) \qquad (2.75)$$

Applying (2.72), expression (II) can be transformed in a similar manner:

$$(II): \quad d_s \left[(\mu^{(1,1)})^2 \right] = d_s \left[\left(\hat{\mu}^{(1,1)} - \mu_u \mu_{u'} \right)^2 \right]$$
$$= 2 \left(d_s \left[\hat{\mu}^{(1,1)} \right] - d_s \left[\mu_u \mu_{u'} \right] \right) \mu^{(1,1)} \qquad (2.76)$$
$$= 2 \left(\mu^{(0,2)} + E\{u \, \partial_u \mathcal{H}\} - \mu_u E\{\partial_u \mathcal{H}\} \right) \mu^{(1,1)}$$

2.2. Charged Particle Dynamics

Using the expressions (I) and (II), equation (2.73) can be written in the final form:

$$d_s \varepsilon_u = \frac{1}{\varepsilon_u} \left[\mu^{(2,0)} \left(E\{u' \partial_u \mathcal{H}\} - \mu_{u'} E\{\partial_u \mathcal{H}\} \right) - \mu^{(1,1)} \left(E\{u \, \partial_u \mathcal{H}\} - \mu_u E\{\partial_u \mathcal{H}\} \right) \right] \quad (2.77)$$

This implies the following conditions in order for the derivative to vanish:

$$E\{u' \partial_u \mathcal{H}\} - \mu_{u'} E\{\partial_u \mathcal{H}\} = \mu^{(1,1)} = \hat{\mu}^{(1,1)} - \mu_u \mu_{u'} \quad (2.78)$$

$$E\{u \, \partial_u \mathcal{H}\} - \mu_u E\{\partial_u \mathcal{H}\} = \mu^{(2,0)} = \hat{\mu}^{(2,0)} - (\mu_u)^2 \quad (2.79)$$

These conditions are fulfilled if:

$$\partial_u \mathcal{H}(u, u') = u \quad (2.80)$$

Referring to the HAMILTON's equations, this condition yields the following theorem:

Theorem 2.2 (Emittance Conservation). *Given a linear* HAMILTONIAN *system defined by:*

$$d_t u = au' \quad (2.81)$$
$$d_t u' = -au \quad (2.82)$$

with $a \in \mathbb{R}$, the emittance ε_u defined by:

$$\varepsilon_u = \left(\sigma_u^2 \cdot \sigma_{u'}^2 - (\sigma_{u,u'})^2 \right)^{1/2} \quad (2.83)$$

is a conserved quantity.

It is, however, emphasized that the projected emittances ε_x and ε_y are in general not conserved quantities due to the coupling of horizontal and vertical motions. The emittance conservation theorem 2.2 specifies conditions for LIOUVILLE's theorem 2.1 to hold true. In particular, it reflects the fact that the emittance is not conserved if non-linear forces act on the beam. This explains the space charge induced emittance growth mentioned at the beginning of this section.

In order to minimize non-linear effects like, e.g., dynamic aperture reduction [141], accelerators are built such that they contain the lowest degree of non-linearity possible. Under linear forces, however, the emittance of a particle beam is conserved. This reveals the practical importance of LIOUVILLE's theorem: The emittance increases from inherent non-linear effects like space charge forces but it cannot be diminished using linear beam line elements.

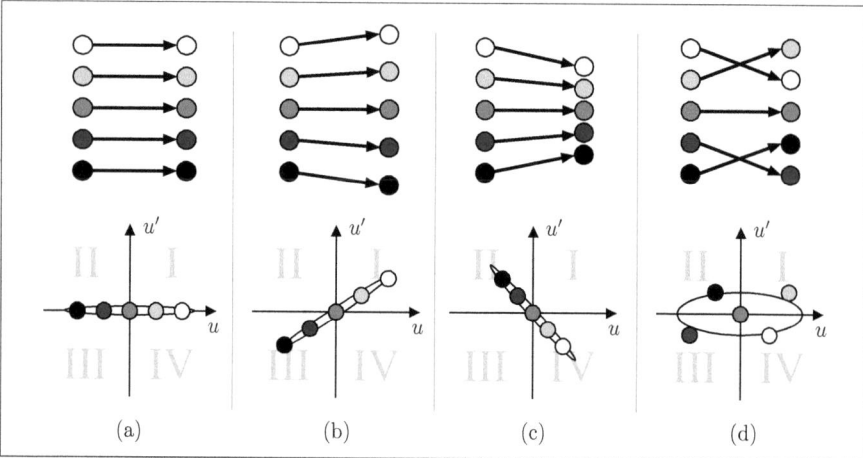

Figure 2.3: Illustration of the connection between the state of a beam and its emittance: The four pictures in the top row depict particle beams consisting of five particles each. The initial positions of the particles in the four cases are identical. Scenario (a) represents a beam with all particles traveling in parallel. In scenario (b) the beam diverges, while in (c) it converges. The beam in scenario (d) is in a mixed state.

The diagrams in the bottom row show the phase space portraits of the respective beams at their final position. The RMS ellipses for the scenarios (a)-(c) all occupy a zero phase space area A. The emittance of the beams vanishes. These particle beams can be transformed into each other by applying linear external forces. In particular, they can be focused to a point-like spot using linear optics. The emittance of beam (d) is non-zero and, therefore, focusing it onto a point is not possible by applying linear forces. This indicates a lower quality of the beam than in (a)-(c).

3. DISCRETE ELECTRODYNAMICS

The first sentence of YEE's seminal paper [144] published in 1966 reads: "Solutions to the time-dependent MAXWELL's equations in general form are unknown except for a few special cases". Moreover, in beam dynamics problems, not only MAXWELL's equations but also the equations of motion have to be solved simultaneously. Nevertheless, it is possible to find an approximate numerical solution of the discretized fundamental equations. The discretization process comprises a spatial and, for transient problems, a temporal discretization step. The former maps the continuous space to a finite set of discrete elements, and the latter introduces discrete instants in time. The approximate solution is determined for every spatial element at each time instant. The research field dealing with the numerical solution of MAXWELL's equations is referred to as *computational electromagnetics (CEM)*.

The set of spatial elements defines a topological structure in space which is called the *computational grid*. Section 3.1 is concerned with specifying properties of the topology of grids that are suitable for the subsequently introduced discretization methods. In addition, terms for describing computational grids and grid refinement techniques are introduced.

From the numerous techniques that are known for discretizing integral or differential equations in space, three methods are introduced in the sections 3.2 to 3.4. These are the *Finite Integration Technique (FIT)*, the *Finite Volume Method (FVM)*, which was combined with the FIT to a hybrid discretization technique, and the *Discontinuous Galerkin Method (DGM)*. Since numerical dispersion is a major concern, the dispersion relations of all schemes are derived. Their analysis is restricted to the propagation of waves along a coordinate, since this case is specially relevant for long particle accelerating structures. The discretization in space leads to semidiscrete formulations. Section 3.5 addresses the discretization of the temporal variable. The key requirements of every numerical scheme are consistency, stability and convergence. In section 3.6 all presented schemes are investigated with regard to these requirements.

Finally, the interaction of the discrete electromagnetic fields and the charged particle dynamics is addressed in section 3.7.

3.1 Discretization of Continuous Space

Computational Grids

The first step in the process of obtaining discrete solutions to MAXWELL's equations is to set up a computational grid $\mathcal{G}$ in the domain of interest. This domain is a subspace of the one-, two- or three-dimensional continuous space. As it was introduced in chapter 2, it is denoted by $\Omega \subset (\mathbb{R}^1, \mathbb{R}^2, \mathbb{R}^3)$. The domain covered by the computational grid is called *computational*

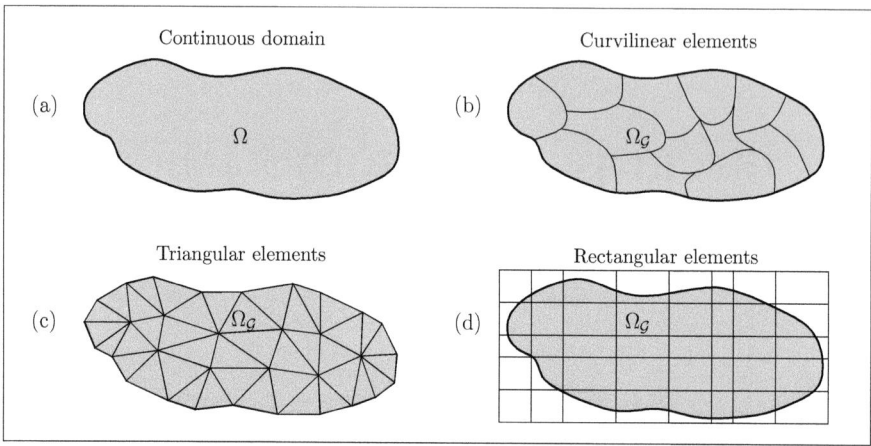

Figure 3.1: Discretization of continuous space: A two-dimensional domain of interest Ω is shown in (a). In order to obtain numerical solutions to MAXWELL's equations, a computational grid $\mathcal{G}$ is set up. In (b) the area is dissected into curvilinear elements, yielding a perfect approximation of Ω. In practice, basic geometrical elements like triangles (c) or rectangles (d) are commonly used. For the discretization of three-dimensional domains volume elements instead of surface elements are employed.

domain $\Omega_{\mathcal{G}}$. This grid $\mathcal{G}$ itself is assembled from individual elements $\mathcal{G}_i$, the grid cells, which are the discrete building blocks of the grid. Depending on the dimensionality of Ω, each cell contains points, edges, faces and a volume.

In order to obtain a grid suitable for the application of the discretization methods presented in the following sections, its elements have to be connected such that they completely cover the domain of interest ($\cup\,\Omega_{\mathcal{G}_i} = \Omega_{\mathcal{G}}$). However, the elements must not overlap ($\cap\,\Omega_{\mathcal{G}_i} = \emptyset$). This general definition allows for more or less arbitrarily shaped elements. For practical reasons (see sections 3.2-3.4) usually basic geometrical shapes like triangles or rectangles in a two-dimensional domain and tetrahedra or hexahedra in three-dimensional domains are employed (see fig. 3.1). Subdividing the domain of interest into elements yields a set of points $\mathbb{P}$, edges $\mathbb{E}$, faces $\mathbb{A}$ and volumes $\mathbb{V}$ (fig. 3.2). Discrete physical quantities can be assigned to any of these. Furthermore, different grids can be used for discretizing different field quantities. Throughout this thesis, *regular grids* composed of hexahedral cells are used. In a three-dimensional regular grid, all faces are shared by exactly two cells, unless the face is located on the domain boundary (see fig. 3.3a).

Grid Refinement

This work deals with techniques for *time-adaptive grid refinement*. Only *conformal refinement* is performed, which preserves the regularity of a computational grid. Non-conformal refinement, on the other hand, introduces *hanging nodes*, which render the grid irregular (see fig. 3.3). Applying conformal refinement to regular grids allows for a continual adjustment of

3.1. Discretization of Continuous Space

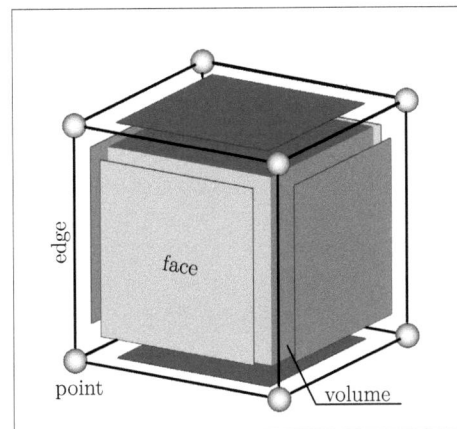

Figure 3.2: Hexahedral grid cell: Illustration of points, edges, and faces as well as the volume of a hexahedral grid cell in an exploded view. Grids assembled from hexahedral cells are employed with the spatial discretization techniques presented in the following sections.

the computational grid while preserving a topological structure that can easily be administrated within a computer program.

For this administration a tree structure was chosen (see fig. 3.4). Starting from an equidistant base grid with a grid step size Δu, grid cells are divided into equally sized halves in every refinement step. Each bisection increments the refinement level L. The local grid step size is

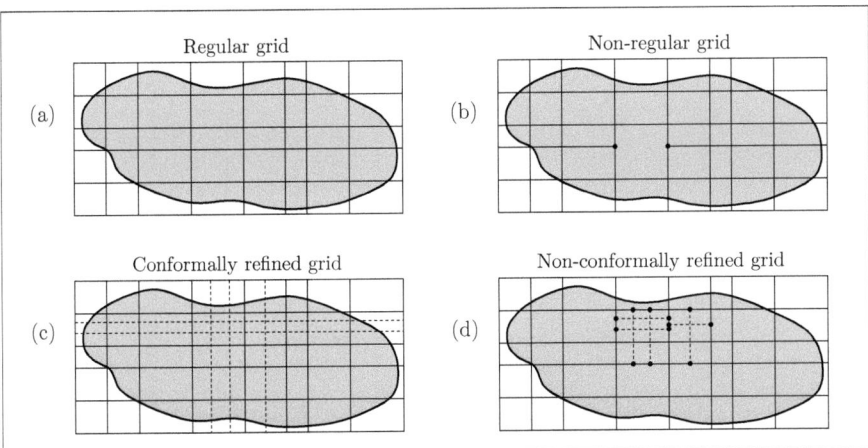

Figure 3.3: Regular grid and conformal refinement: The grid shown in (a) is a regular grid. If one edge is removed, two hanging nodes (indicated as dots) are created, rendering the grid non-regular (b). A conformal refinement technique preserves the regularity of a grid (c), whereas non-conformal refinement destroys the grid regularity.

reduced according to:

$$\Delta u_L = \frac{\Delta u}{2^L} \qquad (3.1)$$

Following this procedure the local grid resolution can be increased while the information about the underlying coarse grid structure is preserved at minimal costs in terms of computer memory. In order to avoid strong variations of the local grid resolution in neighboring areas, the difference in the refinement level L of neighboring cells is limited to one. This ensures a smooth variation of the grid step size.

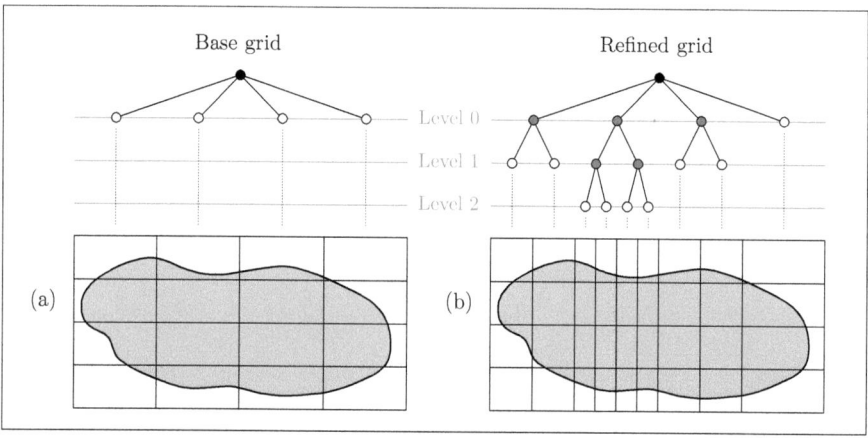

Figure 3.4: Organization of the grid refinement in a tree structure: Starting from a root node (black) the base grid is set up in the form of an N-ary tree. It corresponds to the refinement level zero (a). The grid refinement is organized in binary trees growing from the base grid nodes with every bisection step (b). The leaves of the tree (white) represent the current grid status.

The refinement itself can be adaptive in time. In this case the refined areas and the degree of refinement are not constant in time but subject of adjustment to the current situation. This is illustrated in figure 3.5, where the grid topology at two instants of time is shown. In addition, the tree representations of both grids are shown.

Boundary Approximation Techniques

The domain of interest can include objects with different material characteristics. These objects have to be discretized in order to be represented within the computational domain. In this thesis, the so-called *staircase* approach is applied for this discretization. It is the simplest and fastest among the available options. However, it also has the poorest accuracy. Using a staircase discretization of objects, a grid cell is filled completely with one material. The characteristic material parameters μ, ϵ and κ of the computational cell are assigned according to this material (see figs. 3.4 and 3.7).

More sophisticated methods for the discretization of material interfaces are available. The most well-known of these so-called *boundary conformal* methods are the PFC (partially filled cells)

3.1. Discretization of Continuous Space

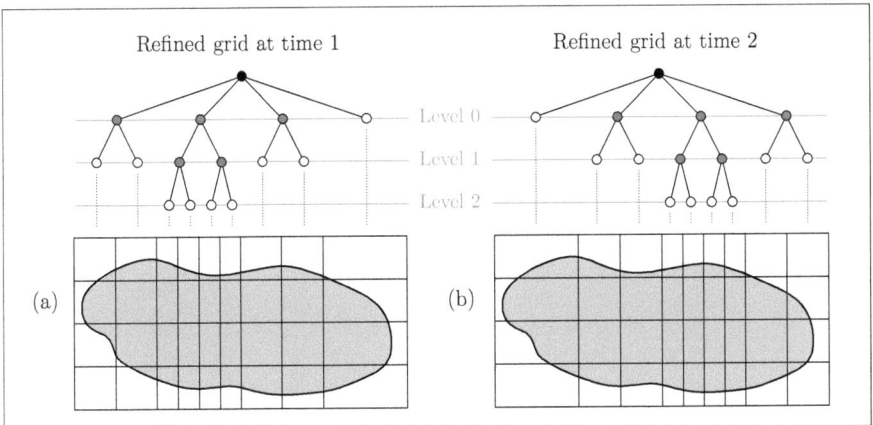

Figure 3.5: Topology of refined grids at two instants in time and their tree representation.

algorithm proposed in [46] and the PBA® (perfect boundary approximation) algorithm implemented in the commercial simulation tool CST STUDIO SUITE™[40]. More recently, the USC (uniformly stable conformal) algorithm and the SC (simplified conformal) scheme were published [147, 148]. Boundary conformal methods yield a more accurate representation of material boundaries, however, at the price of increased computational costs. Every grid adaptation step goes along with a possible reassignment of the characteristic material parameters to adapted cells. Due to the resulting computational overhead, it was refrained from implementing such methods, and staircase fillings are used.

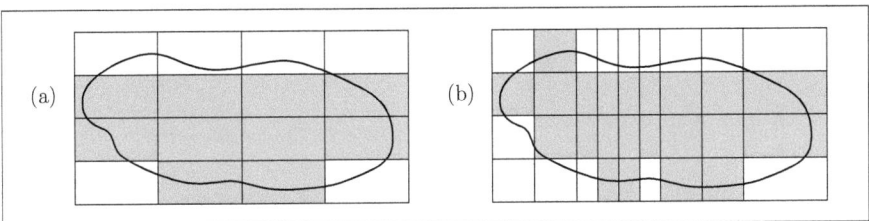

Figure 3.6: Staircase material fillings: In (a) and (b) the respective discrete material distributions for the grids depicted in figure 3.4 are shown. An adaptation of the grid topology goes along with a reassignment of the material properties to the adapted cells.

(a)

(b)

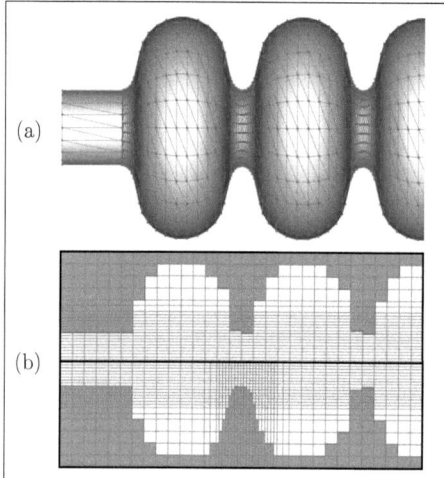

Figure 3.7: Time-adaptive material discretization: In (a) a CAD model of part of an ILC RF accelerating cavity is shown. It is represented by means of a triangulation of its surface (red lines). The resulting discrete material distributions for two instants in time are shown in (b). The upper half corresponds to the distribution within the longitudinally non-refined base grid. In the lower half, the grid is refined in the vicinity of the iris separating the first from the second cell of the cavity. Alongside with the grid refinement goes a more accurate representation of the structure.

3.2 Finite Integration Technique

The Finite Integration Technique (FIT) was introduced by WEILAND in 1977 [136]. Initially, it was applied in the frequency domain for the numerical determination of eigenfrequencies in resonant structures. In the following decades the range of applications was continuously extended. Today, problems in electro- and magnetostatics, frequency domain applications, and time-domain problems ranging from the quasistatic limit to ultra-high frequencies, are successfully solved using the FIT. Since the incorporation of charged particle simulations into the FIT framework [15], the method spans the complete range of CEM problems in accelerator physics.

During the spatial discretization procedure of the FIT, MAXWELL's equations in integral form (2.1)-(2.4) are directly transferred to the grid space. The resulting MAXWELL grid equations, thus, mimic the properties of the equations in continuum.

3.2.1 Semidiscrete Formulation

Local Formulation

The Finite Integration Technique makes use of a pair of grids ($\mathcal{G}$ and $\tilde{\mathcal{G}}$) for the discretization of MAXWELL's equations. In [137] they were named the electric and the magnetic grid. The grids have to fulfill the conditions given in section 3.1. In order to transfer FARADAY's law in integral form (2.2) to the grid space, the loop integral on its left hand side is carried out along the edges of the electric grid $\mathcal{G}$. Integrating the continuous electric field $\vec{E}$ along an edge yields a discrete electric voltage denoted by $\widehat{e}$. Summing up the electric voltages $\widehat{e}_i$ along all edges enclosing a cell face leads to an exact discrete representation of the continuous closed loop integral. Integrating the continuous magnetic flux density $\vec{B}$ over a face yields the discrete magnetic integral flux $\widehat{\widehat{b}}$ (see fig. 3.8). Applying the derivative in time yields a discrete FARADAY's law for each face of the computational grid $\mathcal{G}$ in the form:

$$\sum_i \pm \widehat{e}_i = -\mathrm{d}_t \widehat{\widehat{b}} \qquad (3.2)$$

The orientation of the electric voltages $\widehat{e}_i$ with respect to the loop integral determines their signs (see fig. 3.10). In order to derive this discrete law, nothing more than splitting the loop integral into integrals along the grid edges was done.

For the discretization of AMPÈRE's law, discrete magnetic voltages $\widehat{h}$ are defined as line integrals of the continuous magnetic field $\vec{H}$ along the edges of the magnetic grid $\tilde{\mathcal{G}}$. Consequently, the integration of the continuous electric flux density $\vec{D}$ and the current $\vec{J}$ over a face of $\tilde{\mathcal{G}}$ results in the electric grid flux $\widehat{\widehat{d}}$ and grid current $\widehat{\widehat{j}}$. For every face of $\tilde{\mathcal{G}}$ the discrete AMPÈRE's law reads:

$$\sum_i \pm \widehat{h}_i = \mathrm{d}_t \widehat{\widehat{d}} + \widehat{\widehat{j}} \qquad (3.3)$$

This equation is completely equivalent to the AMPÈRE's law in the integral form (2.1) (cf. figs. 2.1 and 3.8).

The discrete GAUSS' laws are obtained in a straightforward way by summing up the magnetic and electric flux variables $\widehat{\widehat{b}}$ and $\widehat{\widehat{d}}$ defined on all faces of a cell. Fluxes pointing out of the cell

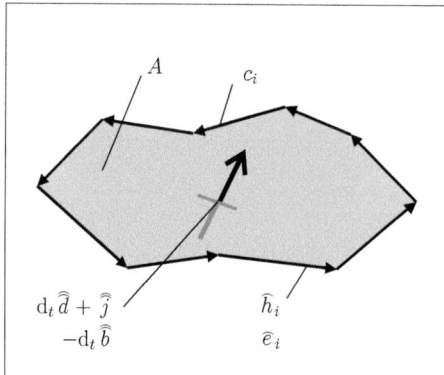

Figure 3.8: Illustration of a possible FIT discretization of AMPÈRE's and FARADAY's law: The equations (3.2) and (3.3) relate the temporal change of the *discrete flux quantities* integrated over the face A to the *sum of the discrete voltage quantities* on the boundary of the face. The face is bounded by a polygon with the segments c_i. Along each segment a line integral is carried out in order to yield the electric and magnetic voltages. This is completely analogous to the continuous case which was illustrated in figure 2.1.

are counted positive, incoming fluxes negative. This yields the discrete equations:

$$\sum_i \pm \widehat{\widehat{b}}_i = 0 \qquad (3.4)$$

$$\sum_i \pm \widehat{\widehat{d}}_i = q \qquad (3.5)$$

for every cell of $\mathcal{G}$ and $\widetilde{\mathcal{G}}$ respectively.

The discrete integral state variables and their definition are summarized in the following:

$$\begin{aligned}
\widehat{e} &:= \int_c \vec{E} \, \mathrm{d}\vec{s} & c &\in \mathbb{E}_{\mathcal{G}} \\
\widehat{h} &:= \int_{\widetilde{c}} \vec{H} \, \mathrm{d}\vec{s} & \widetilde{c} &\in \mathbb{E}_{\widetilde{\mathcal{G}}} \\
\widehat{\widehat{b}} &:= \int_A \vec{B} \, \mathrm{d}\vec{A} & A &\in \mathbb{A}_{\mathcal{G}} \\
\widehat{\widehat{d}} &:= \int_{\widetilde{A}} \vec{D} \, \mathrm{d}\vec{A} & \widetilde{A} &\in \mathbb{A}_{\widetilde{\mathcal{G}}} \\
\widehat{\widehat{j}} &:= \int_{\widetilde{A}} \vec{J} \, \mathrm{d}\vec{A} & \widetilde{A} &\in \mathbb{A}_{\widetilde{\mathcal{G}}} \\
q &:= \int_{\widetilde{V}} \rho \, \mathrm{d}V & \widetilde{V} &\in \mathbb{V}_{\widetilde{\mathcal{G}}}
\end{aligned} \qquad (3.6)$$

In the transformation of MAXWELL's equations to the grid space, no approximations have been introduced. Moreover, no specific relation of the two grids $\mathcal{G}$ and $\widetilde{\mathcal{G}}$ was assumed. However, this relation has to be established for the discretization of the constitutive laws (2.15)-(2.17). At this point it becomes advantageous to employ a staggered pair of dual orthogonal grids. The grid setup proposed by YEE in [144] is a suitable choice for the application of the FIT to problems in three-dimensional Cartesian space. The edges of a primary hexahedral grid $\mathcal{G}$ are aligned with the Cartesian coordinates. Its dual $\widetilde{\mathcal{G}}$ is built such that its grid points $\mathbb{P}_{\widetilde{\mathcal{G}}}$ are co-located with the barycenters of the primary volumes $\mathbb{V}_{\mathcal{G}}$. For pairs of hexahedral grids this already fulfills the orthogonality requirement, i.e., the primary edges penetrate the dual faces perpendicularly and vice versa (see fig. 3.9).

The discretization of the material distribution follows the staircase approach presented in section 3.1. The characteristic material parameters ϵ, μ and κ are assigned to the cells of the

3.2. Finite Integration Technique

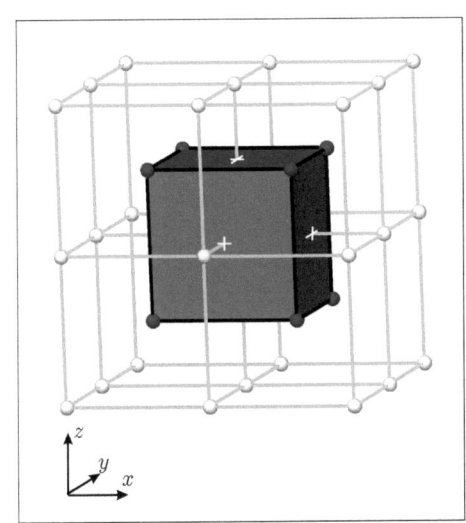

Figure 3.9: Staggered pair of dual orthogonal grids: The FIT deploys a pair of dual orthogonal grids $\mathcal{G}$ and $\widetilde{\mathcal{G}}$. In the following, Cartesian grids are used. They are arranged such that the primary points $\mathbb{P}_\mathcal{G}$ coincide with the barycenters of the dual volumes $\mathbb{V}_{\widetilde{\mathcal{G}}}$ and vice versa. The grids are orthogonal, i.e., edges penetrate the face centers perpendicularly. In this sketch, gray and black can be either the primary or the dual grid.

primary grid $\mathcal{G}$. This ensures an inherent compliance with the continuity conditions of the electromagnetic quantities on material surfaces (see figure 3.10 for an illustration).

After fixing the relation of the two grids $\mathcal{G}$ and $\widetilde{\mathcal{G}}$ as well as the representation of materials in the grid space, the discretization of the constitutive equations can be addressed. The objective is to find a relation between fluxes and voltages, similar to the continuous case. In the FIT these relations are given by:

$$\widehat{\widehat{d}} = M_\epsilon \widehat{e} \qquad (3.7)$$
$$\widehat{\widehat{j}}_\kappa = M_\kappa \widehat{e} \qquad (3.8)$$
$$\widehat{h} = M_{\mu^{-1}} \widehat{\widehat{b}} \qquad (3.9)$$

The discrete material parameters M_ϵ, M_κ and $M_{\mu^{-1}}$ perform a straightforward mapping of the one-dimensional voltages to the two-dimensional fluxes which is given by:

$$M_\epsilon := \bar{\epsilon} \frac{\int_{\widetilde{A}} \mathrm{d}\vec{A}}{\int_c \mathrm{d}\vec{s}}, \qquad M_\kappa := \bar{\kappa} \frac{\int_{\widetilde{A}} \mathrm{d}\vec{A}}{\int_c \mathrm{d}\vec{s}}, \qquad M_{\mu^{-1}} := \overline{\mu^{-1}} \frac{\int_{\widetilde{c}} \mathrm{d}\vec{s}}{\int_A \mathrm{d}\vec{A}} \qquad (3.10)$$

The averaged material characteristics $\bar{\epsilon}, \bar{\kappa}$ and $\overline{\mu^{-1}}$ are defined by [137]:

$$\bar{\epsilon} := \frac{\int_{\widetilde{A}} \epsilon \, \mathrm{d}\vec{A}}{\int_{\widetilde{A}} \mathrm{d}\vec{A}}, \qquad \bar{\kappa} := \frac{\int_{\widetilde{A}} \kappa \, \mathrm{d}\vec{A}}{\int_{\widetilde{A}} \mathrm{d}\vec{A}}, \qquad \overline{\mu^{-1}} := \frac{\int_{\widetilde{c}} 1/\mu \, \mathrm{d}\vec{s}}{\int_{\widetilde{c}} \mathrm{d}\vec{s}} \qquad (3.11)$$

This averaging procedure introduces approximations to the FIT.

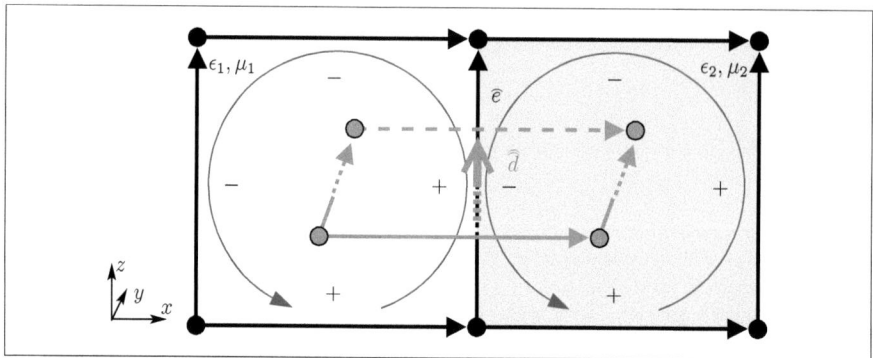

Figure 3.10: The FIT on dual orthogonal grids: In this graph two primary faces (black lines) and one dual face (gray lines) are depicted. Due to the orthogonality of the dual grid, the edges and the associated voltages penetrate the faces perpendicularly in their center. For this grid setup, the conversion of flux values to voltage values and vice versa is given by the equations (3.7)-(3.9).

Since neighboring faces share a common edge and its co-located electric voltage $\widehat{e}$, the boundary condition for the tangential electric field (eqn. (2.29)) is inherently fulfilled if the primary cells are chosen for the material discretization. The same holds true for the boundary condition of the normal magnetic flux density (eqn. (2.28)). The magnetic grid fluxes $\widehat{\widehat{b}}$ are allocated on the primary faces and thus shared by adjacent cells. The signs for the summation of voltages in the discrete FARADAY's and AMPÈRE's law are determined from the edge orientation with respect to the indicated loops.

Matrix Formulation

The N grid points are linearly numbered such that every primary node $P \in \mathbb{P}_\mathcal{G}$ and its dual node $\widetilde{P} \in \mathbb{P}_{\widetilde{\mathcal{G}}}$, located in the barycenter of the associated primary volume $V \in \mathbb{V}_\mathcal{G}$, share a common index. Collecting the discrete state variables defined in equation (3.6) in vectors, yields, e.g., the vector of electric grid voltages:

$$\widehat{\mathbf{e}} := \begin{pmatrix} \widehat{\mathbf{e}}_x \\ \widehat{\mathbf{e}}_y \\ \widehat{\mathbf{e}}_z \end{pmatrix} \quad \text{with} \quad \widehat{\mathbf{e}}_u := \begin{pmatrix} \widehat{e}_u(1) \\ \vdots \\ \widehat{e}_u(N) \end{pmatrix}, \ u \in \{x, y, z\} \tag{3.12}$$

Equivalently, an index triple (i_x, i_y, i_z) can be used for the addressing of the cells and their associated quantities. The vectors $\widehat{\widehat{\mathbf{b}}}$, $\widehat{\mathbf{h}}$, $\widehat{\widehat{\mathbf{d}}}$, $\widehat{\widehat{\mathbf{j}}}$ and $\mathbf{q}$ are defined accordingly. Instead of performing the curl operation locally on each of the $3N$ primary and dual faces (eqns. (3.2), (3.3)), the signs are gathered in the discrete curl matrices $\mathbf{C}$ and $\widetilde{\mathbf{C}}$. This allows for stating the discrete FARADAY's and AMPÈRE's law for the complete computational domain in the compact form:

$$\mathbf{C}\widehat{\mathbf{e}} = -\mathrm{d}_t \widehat{\widehat{\mathbf{b}}} \tag{3.13}$$
$$\widetilde{\mathbf{C}}\widehat{\mathbf{h}} = \mathrm{d}_t \widehat{\widehat{\mathbf{d}}} + \widehat{\widehat{\mathbf{j}}} \tag{3.14}$$

Similarly, discrete divergence matrices $\mathbf{S}$ and $\widetilde{\mathbf{S}}$ can be defined, which represent the signed summation of in- and outgoing fluxes. The discrete GAUSS' laws (3.4) and (3.5) in matrix form

3.2. Finite Integration Technique

read:

$$\mathbf{S}\widehat{\mathbf{b}} = 0 \qquad (3.15)$$

$$\widetilde{\mathbf{S}}\widehat{\widehat{\mathbf{d}}} = \mathbf{q} \qquad (3.16)$$

The equations (3.13)-(3.16) are called the MAXWELL *grid equations* (MGE). The discrete operators $\mathbf{C}$ and $\mathbf{S}$ and their dual counterparts are purely topological. They consist only of the entries +1, 0 and -1, according to the edge and face orientations in the respective loop or surface integral. Introducing the discrete derivative operator $\mathbf{P}_u$ (see [138]) the structure of the discrete curl and divergence operator clearly mimics the continuous operators:

$$\mathbf{C} = \begin{pmatrix} 0 & -\mathbf{P}_z & \mathbf{P}_y \\ \mathbf{P}_z & 0 & -\mathbf{P}_x \\ -\mathbf{P}_y & \mathbf{P}_x & 0 \end{pmatrix} \qquad (3.17)$$

$$\mathbf{S} = \begin{pmatrix} \mathbf{P}_x & \mathbf{P}_y & \mathbf{P}_z \end{pmatrix} \qquad (3.18)$$

Approximations are imposed only by discretizing the constitutive equations. Their matrix formulation reads:

$$\widehat{\widehat{\mathbf{d}}} = \mathbf{M}_\epsilon \widehat{\mathbf{e}} \qquad (3.19)$$

$$\widehat{\widehat{\mathbf{j}}}_\kappa = \mathbf{M}_\kappa \widehat{\mathbf{e}} \qquad (3.20)$$

$$\widehat{\mathbf{h}} = \mathbf{M}_{\mu^{-1}} \widehat{\mathbf{b}} \qquad (3.21)$$

For a dual orthogonal grid doublet, the material matrices $\mathbf{M}_\epsilon, \mathbf{M}_\kappa$ and $\mathbf{M}_{\mu^{-1}}$ are diagonal. They are built from the discrete material parameters defined in (3.10). Exemplary, the composition of $\mathbf{M}_\epsilon$ is given:

$$\mathbf{M}_\epsilon := \mathrm{diag}\left(\mathbf{M}_{\epsilon,x}, \mathbf{M}_{\epsilon,y}, \mathbf{M}_{\epsilon,z}\right) \quad \text{with} \quad \mathbf{M}_{\epsilon,u} := \mathrm{diag}(M_{\epsilon,u}(i)),\ i = 1..N, u \in \{x,y,z\} \qquad (3.22)$$

The direction of u corresponds to the direction the integration path c in equation (3.10).

3.2.2 Properties of the Finite Integration Technique

In this section some algebraic properties of the FIT are highlighted. Shorthand, one can say that due to the immediate transfer of MAXWELL's equations to the grid space, the properties of the equations in continuum and their solutions are still exhibited in the MAXWELL grid equations and their discrete solutions. The matrix notation introduced in the last section, is not only an elegant way of expressing the discrete equations, but it also allows for rather simple proofs of, e.g., energy and charge conservation of the discrete equations.

The compliance with the boundary conditions (2.27)-(2.30) is a major advantage of the FIT over *Finite Element Methods* (FEM) employing nodal basis functions [150], the Finite Volume Method (presented in sec. 3.3) and the Discontinuous Galerkin Method (presented in sec. 3.4). Spurious modes are excluded from the space of solutions [33, 139].

Conservation Properties

In the coupled simulation of MAXWELL's equations and charged particle dynamics the conservation of charge is an important property. In FIT notation, the vector calculus identity $\nabla \cdot (\nabla \times \vec{a}) = 0$ reads [33]:

$$\mathbf{SC} = \mathbf{0}, \quad \widetilde{\mathbf{S}}\widetilde{\mathbf{C}} = \mathbf{0} \tag{3.23}$$

The proof of the conservation of charge in the grid space is coherent with the continuous case shown in section 2.1.2 using discrete operators:

$$\widetilde{\mathbf{S}}\widetilde{\mathbf{C}}\widehat{\mathbf{h}} = d_t \widetilde{\mathbf{S}}\widehat{\mathbf{d}} + \widetilde{\mathbf{S}}\widehat{\widetilde{\mathbf{j}}} \tag{3.24}$$

$$\mathbf{0} = d_t \mathbf{q} + \widetilde{\mathbf{S}}\widehat{\widetilde{\mathbf{j}}} \tag{3.25}$$

Since only topological operators are involved, the discrete continuity equation (3.25) contains no approximations. The charge $\mathbf{q}$ is, thus, an exactly conserved quantity. In section 3.7 the continuity equation is used to determine the convective grid currents induced by moving charges.

In order to demonstrate the conservation of energy, the vectors of voltages $\widehat{\mathbf{h}}$ and $\overline{\mathbf{e}}$ are collected into a vector $\mathbf{u}$:

$$\mathbf{u} = \begin{pmatrix} \widehat{\mathbf{h}} \\ \overline{\mathbf{e}} \end{pmatrix} \tag{3.26}$$

The time-dependent energy $\mathcal{W}$ of the spatially discrete electromagnetic fields is defined by means of the energy norm:

$$\mathcal{W}(t) := \|\mathbf{u}\|_E = \frac{1}{2}\left(\widehat{\mathbf{h}}^T \mathbf{M}_{\mu^{-1}}^{-1} \widehat{\mathbf{h}} + \overline{\mathbf{e}}^T \mathbf{M}_\epsilon \overline{\mathbf{e}}\right) \tag{3.27}$$

which is analogous to the continuous case (sec. 2.1.4). The definition (3.27) is a norm because the diagonal material matrices $\mathbf{M}_{\mu^{-1}}$ and $\mathbf{M}_\epsilon$ are positive definite. In the FIT a duality relation for the primary and dual curl operator:

$$\mathbf{C} = \widetilde{\mathbf{C}}^T \tag{3.28}$$

is given for the above stated dual orthogonal grid doublet. This relation can equally be expressed by $\mathbf{P} = -\widetilde{\mathbf{P}}^T$ using a dual derivative operator. Using this duality, it is shown in [33] that for a computational domain enclosed by perfectly conducting boundaries the change of the discrete electromagnetic energy with respect to time is given by:

$$d_t \mathcal{W} = -\overline{\mathbf{e}}^T \widehat{\widetilde{\mathbf{j}}} \tag{3.29}$$

This is again in a full analogy to the continuous case given in equation (2.22). In the absence of currents

$$d_t \mathcal{W}(t) = 0 \tag{3.30}$$

holds true.

Another approach to derive the conservation of energy starts from the formulation of the MAXWELL grid equations as one system of ordinary differential equations (ODE):

$$d_t \mathbf{u} = \underbrace{\begin{pmatrix} \mathbf{0} & -\mathbf{M}_{\mu^{-1}}\mathbf{C} \\ \mathbf{M}_\epsilon^{-1}\widetilde{\mathbf{C}} & \mathbf{0} \end{pmatrix}}_{\mathbf{A}} \mathbf{u} \tag{3.31}$$

3.2. Finite Integration Technique

Here, conductive materials as well as currents have been neglected. Due to the positive definiteness of the material matrices, a similarity transformation of the *system matrix* **A** can be performed [114]. The transformed system reads:

$$d_t \mathbf{u}' = \mathbf{A}' \mathbf{u}' \tag{3.32}$$

with

$$\mathbf{A}' = \begin{pmatrix} 0 & -\mathbf{M}_{\mu^{-1}}^{1/2} \mathbf{C} \mathbf{M}_\epsilon^{-1/2} \\ \mathbf{M}_\epsilon^{-1/2} \tilde{\mathbf{C}} \mathbf{M}_{\mu^{-1}}^{1/2} & 0 \end{pmatrix} \tag{3.33}$$

and

$$\mathbf{u}' = \begin{pmatrix} \widehat{\mathbf{h}}' \\ \widehat{\mathbf{e}}' \end{pmatrix} \quad \text{with} \quad \widehat{\mathbf{e}}' = \mathbf{M}_\epsilon^{1/2} \widehat{\mathbf{e}}, \quad \widehat{\mathbf{h}}' = \mathbf{M}_{\mu^{-1}}^{1/2} \widehat{\mathbf{h}} \tag{3.34}$$

Due to the duality relation (3.28), the matrix **A**′ is skew-symmetric. Therefore, its eigenvalues, and also the eigenvalues of the system matrix **A**, are either zero or pairwise conjugate and purely imaginary. Thus, the semidiscrete equations above preserve the HAMILTONIAN structure of the MAXWELL's equations in continuum which guarantees the energy to be conserved [146].

Every type of grid refinement applied within the FIT framework necessarily has to maintain its algebraic properties. In case of a violation of (3.23) or (3.28), the charge or energy conserving property of the FIT is lost. For conformal grid refinements, this prerequisite is automatically fulfilled since the dual orthogonal structure of the grid is kept at every time.

Dispersion Properties

In the following, the grid dispersion properties of the FIT are addressed. In accordance with section 2.1.6, the combination of FARADAY's and AMPÈRE's law yields the *discrete homogeneous wave equation*:

$$\tilde{\mathbf{C}} \mathbf{M}_{\mu^{-1}} \mathbf{C} \widehat{\mathbf{e}} + \mathbf{M}_\epsilon d_t^2 \widehat{\mathbf{e}} = 0 \tag{3.35}$$

for the electric field in lossless media where, again, currents were neglected (cf. eqn. (2.32)). For the time-harmonic case this equation can be written in the form:

$$\mathbf{M}_\epsilon^{-1} \tilde{\mathbf{C}} \mathbf{M}_{\mu^{-1}} \mathbf{C} \widehat{\mathbf{e}} = \omega^2 \widehat{\mathbf{e}} \tag{3.36}$$

which is called the *curl-curl eigenvalue equation* of the electric field. In section 2.1.6 it was shown that the solutions of the wave equation (2.32) are superpositions of plane waves of the form:

$$\vec{E} \sim e^{i(\vec{k} \cdot \vec{r} - \omega t)} \tag{3.37}$$

Using a plane wave ansatz, a VON NEUMANN analysis [39, 90] is carried out. In this analysis an infinite grid of identical cells is assumed. It is, hence, sufficient to consider one representative cell. The grid voltages of its neighboring cells are obtained by the multiplication with a phase factor:

$$\widehat{e}_u(i_u + 1) = \widehat{e}_u(i_u) \cdot e^{-i k_u \Delta u} \tag{3.38}$$

This allows for reducing (3.36) to a local 3 × 3 eigenvalue equation. Its solutions, λ_A, are real valued and allow for establishing the semidiscrete *grid dispersion relation*. It reads:

$$\sum_u \left[\left(\frac{\sin\left(\frac{k_u \Delta u}{2}\right)}{\frac{\Delta u}{2}} \right)^2 \right] = \left(\frac{\omega}{c} \right)^2 \quad \text{with} \quad u \in \{x, y, z\} \tag{3.39}$$

If the plane waves are assumed to propagate along an axis of the coordinate system, all but one term of the sum are equal to zero. Then, the relation can be written as:

$$2 \sin\left(\frac{k_u \Delta u}{2}\right) = \Delta u \frac{\omega}{c} \qquad (3.40)$$

In figure 3.11 this is plotted as a function of the phase advance per cell, β, along the coordinate u:

$$\beta_u := k_u \Delta u \qquad (3.41)$$

The graph shows that all discrete waves show a phase lag compared to the continuous case independently of the grid resolution. However, if the grid step size is very small ($\beta \to 0$), the physical phase speed is recovered. From the graph it is clear that dispersion errors are largest for short wave lengths. They can be reduced by applying local grid refinement in such regions of high-frequency fields.

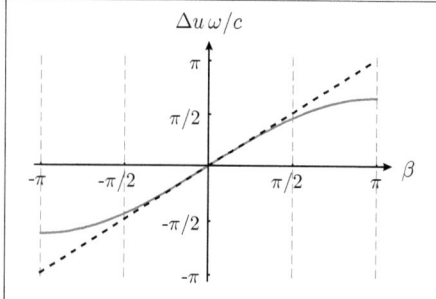

Figure 3.11: FIT grid dispersion relation: The plot shows the numerical phase advance of plane waves in the grid space traveling along a coordinate axis u in comparison to the phase advance in continuous space (dashed line). Discrete waves of all frequencies show a phase lag in comparison to the continuous case. In the long wave limit ($\beta \to 0$) the physical phase velocity is recovered.

It is remarked that for transient problems, the FIT contains YEE's algorithm as a special case. The main issues of the both were also mentioned by YEE already: the incorporation of curved material surfaces and the treatment of electrically large bodies. The former can be considered solved since the publication of computationally inexpensive schemes for a conformal treatment of boundaries [148]. At least for some cases, the latter can be tackled by time-adaptive grid refinement techniques. One of these cases is the simulation of highly localized particle bunches, which is covered in this thesis.

3.3 Finite Integration-Finite Volume Hybrid Method

The preceding section finished with a discussion of the dispersion properties of the Finite Integration Technique. There, it was shown that all waves traveling in the grid space lag behind their continuous counterparts. The phase lag increases as the grid resolution is decreased. In this section, Finite Volume Methods (FVM), which are known for their well-behaved dispersion properties, are investigated. This motivates the construction of a FI-FV mixed scheme with improved dispersion properties along one coordinate.

In the context of Finite Volume Methods, the discrete state variables are obtained via the integration of the physical quantities over the volume of every grid cell. The coupling of the individual cells is carried out via so-called *numerical fluxes*. A large number of flux definitions can be found in the literature. Choosing a numerical flux yields a specific FVM.

The mathematical foundations of the FVM were established by LAX and GODUNOV starting about 1950. More detailed information can be found in the monographs by LEVEQUE [83, 84] and a series of publications by VAN LEER [131]. Finite Volume Methods are widely used in computational fluid dynamics. In CEM they gained some popularity over the last decade, mainly because the methods allow for a simple and computationally inexpensive use of unstructured grids in time-domain simulations (see e.g. [91]). However, depending on the formulation, a non-physical dissipation of energy is introduced. Unlike the FIT, FV methods do not inherently fulfill the boundary conditions imposed by the MAXWELL's equations.

In order to illustrate the reduced dispersion errors of the method, the propagation of a rectangular wave packet using a coarse grid is investigated in a one-dimensional simulation. The initial state of the packet is plotted in figure 3.12a. The orientations of the electric and magnetic field are chosen such that the wave vector $\vec{k}$ points in the positive z-direction. The domain is bounded with a PEC material. The propagation was simulated with the FIT and a FV method using the same spatial grid resolution. The results are illustrated in figure 3.12b (FVM) and 3.12c (FIT). The improved dispersion properties make FV methods an appropriate choice for the handling of the very high-frequency fields, excited by short relativistic bunches.

The starting point for the derivation of the hybrid FI-FV scheme is a separation of the continuous curl operator according to:

$$\nabla \times = \begin{pmatrix} 0 & -\partial/\partial z & \partial/\partial y \\ \partial/\partial z & 0 & -\partial/\partial x \\ -\partial/\partial y & \partial/\partial x & 0 \end{pmatrix} = \underbrace{\begin{pmatrix} 0 & 0 & \partial/\partial y \\ 0 & 0 & -\partial/\partial x \\ -\partial/\partial y & \partial/\partial x & 0 \end{pmatrix}}_{\hat{=} \mathbf{C}^{\text{2D-FIT}}} + \underbrace{\begin{pmatrix} 0 & -\partial/\partial z & 0 \\ \partial/\partial z & 0 & 0 \\ 0 & 0 & 0 \end{pmatrix}}_{\hat{=} \mathbf{C}^{\text{1D-FV}}} \qquad (3.42)$$

$$\underbrace{\phantom{\begin{pmatrix} 0 & -\partial/\partial z & \partial/\partial y \\ \partial/\partial z & 0 & -\partial/\partial x \\ -\partial/\partial y & \partial/\partial x & 0 \end{pmatrix}}}_{\hat{=} \mathbf{C}^{\text{3D}}}$$

The resulting two-dimensional problem is handled using the FIT, while a Finite Volume Method is applied to the one-dimensional problem. This is illustrated in figure 3.13. The above idea of splitting the curl operation was proposed in [80]. There, however, the FIT discretization is exclusively used. In the following, the FV formulation is given for a one-dimensional problem involving the partial derivative ∂_z along the z-coordinate and the electromagnetic flux components B_x, B_y, D_x, D_y only. However, at the end of this section, also the system matrix for three-dimensional hybrid method will be stated.

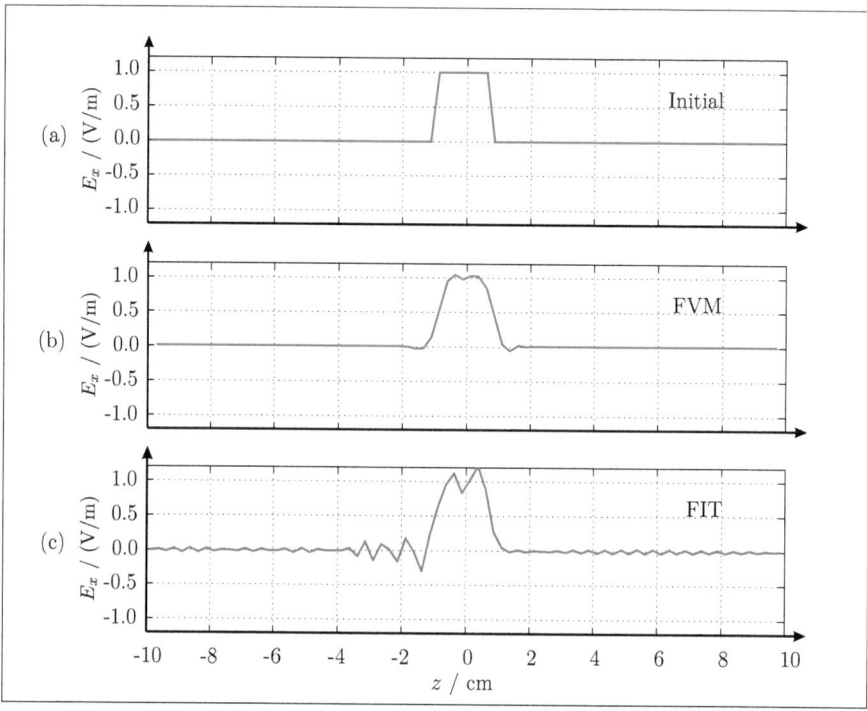

Figure 3.12: Dispersion of FIT and FV: A one-dimensional CAUCHY problem is chosen for an illustration of the dispersion properties of the Finite Integration Technique and a Finite Volume Method. A rectangular field distribution is imposed as initial conditions on a coarse grid (a). The propagation of this field distribution is simulated using identical grids. In the plots of the final field distribution, the superior dispersion properties of the FVM (b) over the FIT (c) are visible.

3.3.1 Semidiscrete Finite Volume Formulation in 1D

Local Formulation

FARADAY's and AMPÈRE's law are cast into the conservative form [18]. For the considered one-dimensional case, this yields:

$$\partial_t \vec{U} + \partial_z \mathcal{F}(\vec{U}) = 0, \qquad \vec{U} = \begin{pmatrix} \vec{B}^{1D} \\ \vec{D}^{1D} \end{pmatrix} \tag{3.43}$$

with $\vec{B}^{1D} = (B_x, B_y)^T$ and $\vec{D}^{1D} = (D_x, D_y)^T$ and the *flux*:

$$\mathcal{F}(\vec{U}) = (-D_y/\epsilon, D_x/\epsilon, B_y/\mu, -B_x/\mu)^T = (-E_y, E_x, H_y, -H_x)^T \tag{3.44}$$

3.3. Finite Integration-Finite Volume Hybrid Method

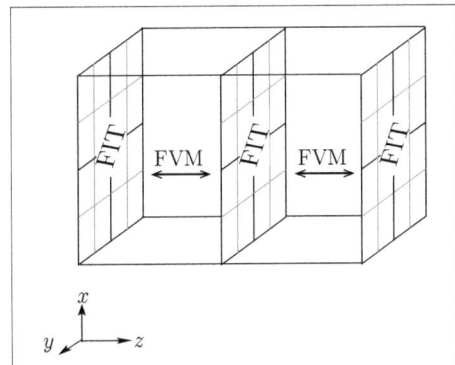

Figure 3.13: Illustration of the FI-FV Method in space: The FIT is applied in parallel $x - y$-planes of the computational grid. The FVM is applied in the z-direction for connecting these planes.

Integrating (3.43) over a finite interval in space yields:

$$d_t \int_{\Delta z} \vec{U} \, dz + \sum_i \mathcal{F}^i(\vec{U}^*) = 0 \quad \text{with} \quad i = 1, 2 \quad (3.45)$$

where the divergence theorem was applied. The value of $\vec{U}$ at the interval boundaries is denoted by $\vec{U}^*$. Hence, the temporal change of the electromagnetic quantities in the interval equals the sum of the boundary fluxes $\mathcal{F}^i(\vec{U}^*)$. The DoF in the framework of Finite Volume Methods are averaged integral state variables given by:

$$\vec{u}(i_z) := \frac{1}{\Delta z(i_z)} \int_{\Delta z(i_z)} \vec{U} \, dz \quad \text{with} \quad \vec{u}(i_z) = (b_x, b_y, d_x, d_y)^\mathrm{T} \quad (3.46)$$

with $\Delta z(i_z)$ the width of the interval i_z. Inserting this into (3.45) yields the local formulation:

$$d_t \vec{u}(i_z) + \frac{1}{\Delta z(i_z)} \left[\mathcal{F}\left(\vec{U}^*(z(i_z + 1/2))\right) - \mathcal{F}\left(\vec{U}^*(z(i_z - 1/2))\right) \right] = 0, \quad \forall i_z = [1..N_z] \quad (3.47)$$

with

$$z(i_z \pm 1/2) = z(i_z) \pm \frac{\Delta z(i_z)}{2} \quad (3.48)$$

where $z(i_z)$ are the interval midpoints. Due to the definition of the DoF given in (3.46), the value at interval boundaries is ambiguous. There exist two values, $\vec{U}^+$ and $\vec{U}^-$, one on the positive and one on the negative side of the boundary with respect to the z-coordinate. Two options for obtaining a unique boundary ([77, 83]) value are the central value:

$$\begin{aligned} E_x^* &:= \tfrac{1}{2}(E_x^+ + E_x^-), & H_x^* &:= \tfrac{1}{2}(H_x^+ + H_x^-) \\ E_y^* &:= \tfrac{1}{2}(E_y^+ + E_y^-), & H_y^* &:= \tfrac{1}{2}(H_y^+ + H_y^-) \end{aligned} \quad (3.49)$$

and the upwind value:

$$\begin{aligned} E_x^* &:= \frac{\left(Y^+ E_x^+ - H_y^+\right) + \left(Y^- E_x^- + H_y^-\right)}{Y^+ + Y^-}, & H_x^* &:= \frac{\left(Z^+ H_x^+ + E_y^+\right) + \left(Z^- H_x^- - E_y^-\right)}{Z^+ + Z^-} \\ E_y^* &:= \frac{\left(Y^+ E_y^+ + H_x^+\right) + \left(Y^- E_y^- - H_x^-\right)}{Y^+ + Y^-}, & H_y^* &:= \frac{\left(Z^+ H_y^+ - E_x^+\right) + \left(Z^- H_y^- + E_x^-\right)}{Z^+ + Z^-} \end{aligned} \quad (3.50)$$

with
$$Z^{\pm} := \sqrt{\frac{\epsilon^{\pm}}{\mu^{\pm}}} \quad \text{and} \quad Y^{\pm} := \frac{1}{Z^{\pm}} \tag{3.51}$$

The upwind value is the exact solution of the MAXWELL's equations for piecewise constant initial data after an infinitesimal time span. This referred to as the RIEMANNIAN problem [77]. Its solution is outlined in the following.

Locally, equation (3.43) can be rewritten as:

$$\partial_t \begin{pmatrix} \vec{B}^{1D} \\ \vec{D}^{1D} \end{pmatrix} + \partial_z \mathbf{F} \begin{pmatrix} \vec{B}^{1D} \\ \vec{D}^{1D} \end{pmatrix} = 0 \tag{3.52}$$

with

$$\mathbf{F} = \begin{pmatrix} 0 & 0 & 0 & -1/\mu \\ 0 & 0 & 1/\mu & 0 \\ 0 & 1/\epsilon & 0 & 0 \\ -1/\epsilon & 0 & 0 & 0 \end{pmatrix} \tag{3.53}$$

The eigenvalues of $\mathbf{F}$ are:
$$\text{eig}(\mathbf{F}) = \{-c, -c, c, c\} \tag{3.54}$$

which are the characteristics of the scalar homogeneous wave equation in one dimension. They correspond to the two possible propagation directions for each of the two polarizations of the plane wave solution. The matrix $\mathbf{F}$ can be split such that:

$$\mathbf{F} = \mathbf{F}^+ + \mathbf{F}^- \tag{3.55}$$

with

$$\mathbf{F}^+ = \begin{pmatrix} c/2 & 0 & 0 & -1/2\mu \\ 0 & c/2 & 1/2\mu & 0 \\ 0 & 1/2\epsilon & c/2 & 0 \\ -1/2\epsilon & 0 & 0 & c/2 \end{pmatrix} \quad \text{and} \quad \mathbf{F}^- = \begin{pmatrix} -c/2 & 0 & 0 & -1/2\mu \\ 0 & -c/2 & 1/2\mu & 0 \\ 0 & 1/2\epsilon & -c/2 & 0 \\ -1/2\epsilon & 0 & 0 & -c/2 \end{pmatrix} \tag{3.56}$$

The eigenvalues of $\mathbf{F}^+$ and $\mathbf{F}^-$ read:

$$\text{eig}(\mathbf{F}^+) = \{0, 0, c, c\} \quad \text{and} \quad \text{eig}(\mathbf{F}^-) = \{-c, -c, 0, 0\} \tag{3.57}$$

Thus, $\mathbf{F}^+$ and $\mathbf{F}^-$ describe propagations along the characteristics in the positive and negative z-direction only. Using (3.52), (3.55) and the relations:

$$c = \frac{Z}{\epsilon} \quad \text{and} \quad c = \frac{Y}{\mu} \tag{3.58}$$

the upwind values (3.50) after an infinitesimal time span are found. The semi-discrete FV formulation reads:

$$d_t \vec{u}(i_z) + \frac{1}{\Delta z(i_z)} \left[\vec{f}(i_z + 1/2) - \vec{f}(i_z - 1/2) \right] = 0, \quad \forall i_z = [1..N_z] \tag{3.59}$$

with the vector of *numerical fluxes*:

$$\vec{f}(i_z \pm 1/2) = \Big(-E_y^*(i_z \pm 1/2), E_x^*(i_z \pm 1/2), H_y^*(i_z \pm 1/2), -H_x^*(i_z \pm 1/2) \Big)^{\mathrm{T}} \tag{3.60}$$

3.3. Finite Integration-Finite Volume Hybrid Method

If the boundary values (3.49) are used, the flux is called central while it is called upwind flux if the definition (3.50) is used.

It remains to address the determination of the values on the positive and negative side of an interval boundary. In order to reconstruct the electromagnetic field within the interval i_z, we make the linear ansatz:

$$E_x(i_z) = \frac{d_x(i_z)}{\epsilon}(z - z(i_z)) s(E_x) \qquad (3.61)$$

for the E_x component and all other components likewise. The *slope* s is an approximation of arbitrary order to the derivative of the respective physical quantity:

$$s(E_x) \simeq \partial_z E_x \qquad (3.62)$$

Different FV methods use different slope definitions. For GODUNOV's method, it is given by:

$$s^G(E_x) := 0 \qquad (3.63)$$

yielding a piecewise constant reconstruction within every cell. Other methods take a shift in space from the cell center to the interface into account. In the LAX-WENDROFF scheme, the slope reads[1]:

$$s^{LW}(E_x) := \begin{cases} \frac{e_x(i_z+1)-e_x(i_z)}{z(i_z+1)-z(i_z)} & \text{if } z > z(i_z) \\ \frac{e_x(i_z-1)-e_x(i_z)}{z(i_z)-z(i_z-1)} & \text{if } z < z(i_z) \end{cases} \qquad (3.64)$$

In FROMM's scheme, central differences are used for the approximation:

$$s^F(E_x) := \frac{e_x(i_z+1) - e_x(i_z-1)}{z(i_z+1) - z(i_z-1)} \qquad (3.65)$$

The boundary values of E_x are obtained by:

$$E_x(i_z)^\pm = \frac{d_x(i_z)}{\epsilon} \mp \frac{\Delta z(i_z)}{2} s(E_x) \qquad (3.66)$$

Matrix Formulation

Arranging the semidiscrete FV formulation in the form of a system of ODEs yields (cf. eqn. (3.31)):

$$d_t \mathbf{u} = \mathbf{A}\mathbf{u} \qquad (3.67)$$

The system matrix employing central fluxes reads:

$$\mathbf{A} = \begin{pmatrix} 0 & 0 & 0 & \mathbf{D}_\mu^{-1}\mathbf{P}_z \\ 0 & 0 & -\mathbf{D}_\mu^{-1}\mathbf{P}_z & 0 \\ 0 & \mathbf{D}_\epsilon^{-1}\mathbf{P}_z & 0 & 0 \\ -\mathbf{D}_\epsilon^{-1}\mathbf{P}_z & 0 & 0 & 0 \end{pmatrix} \qquad (3.68)$$

where $\mathbf{P}_z$ is the discrete FV derivative operator [79] and $\mathbf{D}_\mu = \text{diag}(\mu(i_z))$, $\mathbf{D}_\epsilon = \text{diag}(\epsilon(i_z))$ are the material matrices. It allows for identifying the discrete FV curl operator $\mathbf{C}$ for the one-dimensional case under consideration:

$$\mathbf{C} = \begin{pmatrix} 0 & -\mathbf{P}_z \\ \mathbf{P}_z & 0 \end{pmatrix} \qquad (3.69)$$

---

[1] Here the slope is in principal a function of E_x and z: $s^{LW}(E_x, z)$.

In this form the system matrices according to the FIT and the FVM are formally identical. Both of them can be transformed to a skew-symmetric form. Their eigenvalues are pairwise purely imaginary. Thus, they can be symplectically integrated in time (cf. sec. 3.2.2).

The classical FV formulation, however, involves upwind fluxes. In this case the system matrix reads:

$$\mathbf{A} = \begin{pmatrix} \mathbf{U}_h & 0 & 0 & \mathbf{D}_\mu^{-1}\mathbf{P}_z \\ 0 & \mathbf{U}_h & -\mathbf{D}_\mu^{-1}\mathbf{P}_z & 0 \\ 0 & \mathbf{D}_\epsilon^{-1}\mathbf{P}_z & \mathbf{U}_e & 0 \\ -\mathbf{D}_\epsilon^{-1}\mathbf{P}_z & 0 & 0 & \mathbf{U}_e \end{pmatrix} \quad (3.70)$$

with the upwind operators $\mathbf{U}_h$ and $\mathbf{U}_e$ [79].

3.3.2 Properties of Finite Volume Methods in 1D

The accuracy of the spatial discretization of FV methods depends on the reconstruction step using the slope s. For the GODUNOV, LAX-WENDROFF and FROMM scheme, the accuracy of the approximation is of the order zero, one and two. Expressions with a higher accuracy can be derived by including more derivatives. This, in turn, leads to larger stencils which cause problems at material boundaries.

The grid dispersion relations can be derived by means of a VON NEUMANN analysis assuming time-harmonic plane waves. The procedure completely agrees with the analysis carried out for the FIT in section 3.2.2. Due to the upwind blocks, the FV system matrix does not exhibit an antisymmetric structure. The resulting local eigenvalue problem, therefore, has general complex solutions $\lambda_A \in \mathbb{C}$. They read:

$$\frac{e^{i\,\mathrm{Re}[\lambda_A]} \cdot e^{-\mathrm{Im}[\lambda_A]}}{\Delta z} = \frac{\omega}{c}$$
$$\Phi(k\Delta z) \cdot A(k\Delta z) = \Delta z \frac{\omega}{c} \quad (3.71)$$

with the phase function Φ and the amplification function A:

$$\Phi(k\Delta z) = e^{i\,\mathrm{Re}[\lambda_A]} \quad (3.72)$$
$$A(k\Delta z) = e^{-\mathrm{Im}[\lambda_A]} \quad (3.73)$$

For the three methods, they read:

- GODUNOV:

$$\Phi^\mathrm{G}(k\Delta z) = \sin(k\Delta z) \quad (3.74)$$
$$A^\mathrm{G}(k\Delta z) = \exp\left(\cos(k\Delta z) - 1\right) \quad (3.75)$$

- LAX-WENDROFF:

$$\Phi^\mathrm{LW}(k\Delta z) = \left(2 - \cos(k\Delta z)\right)\sin(k\Delta z) \quad (3.76)$$
$$A^\mathrm{LW}(k\Delta z) = \exp\left(4 \cdot \sin^4\left(\frac{k\Delta z}{2}\right)\right) \quad (3.77)$$

3.3. Finite Integration-Finite Volume Hybrid Method

- FROMM:

$$\Phi^F(k\Delta z) = \left(\frac{3 - \cos(k\Delta z)}{2}\right)\sin(k\Delta z) \qquad (3.78)$$

$$A^F(k\Delta z) = \exp\left(2 \cdot \sin^4\left(\frac{k\Delta z}{2}\right)\right) \qquad (3.79)$$

They are plotted in figure 3.15. Instead of the amplification, the damping factor, $1 - A$, is shown. Comparing the semidiscrete dispersion relations of the FIT (3.39) and GODUNOV's method (3.74), it is evident that GODUNOV's method requires twice the grid resolution in order to yield the same phase accuracy. This is a direct consequence of the staggered pair of grids employed by the FIT. It can be concluded that the LAX-WENDROFF or FROMM method should be chosen in order to benefit from enhanced dispersion properties. The example presented in figure 3.12 was simulated using FROMM's method.

3.3.3 The Hybrid Scheme in 3D

Unlike the FIT, the FVM does not apply a staggered grid doublet. All quantities are discretized on the same grid. The three-dimensional computational grid, thus, consists of dual orthogonal grid layers in the $x - y$-planes which are connected in a non-staggered fashion along the z-coordinate. In figure 3.14 a detailed sketch is shown. The primary FIT grid is indicated in black, the dual in gray. The staggered FIT grid layers are arranged in the center of the cells along the z-axis. The outline of groups of four cells is indicated with black dashed boxes. Within this setup, the FVM is applied along the dash-dotted z-directed lines.

Coupling of the DoF

The electric and magnetic field DoF of the FIT and FVM are defined differently. The former are integral state variables obtained by an integration over edges and faces (eqn. (3.6)). The latter are integral state variable obtained by integrating over a cell and averaging by its size. Hence, their coupling has to be addressed. In the figure 3.14, the primary and respective dual nodes marked as dots are indexed by (i_x, i_y, i_z) and $(i_x, i_y, i_z + 1)$. The red arrows indicate the electric grid voltages $\widehat{e}_x(i_x, i_y, i_z)$, $\widehat{e}_y(i_x, i_y, i_z)$ and $\widehat{e}_x(i_x, i_y, i_z+1)$, $\widehat{e}_y(i_x, i_y, i_z+1)$. The green arrows indicate the respective magnetic voltages $\widehat{h}_x(\widetilde{i}_x, \widetilde{i}_y, i_z)$, $\widehat{h}_y(\widetilde{i}_x, \widetilde{i}_y, i_z)$ and $\widehat{h}_x(\widetilde{i}_x, \widetilde{i}_y, i_z+1)$, $\widehat{h}_y(\widetilde{i}_x, \widetilde{i}_y, i_z+1)$. For simplicity, the indices i_x and i_y are omitted in the following. The notation is, thus, $\widehat{e}_x(i_z)$ etc. for the FIT variables and $d_x(i_z)$ etc. for the FV variables.

In order to obtain a sampled field value in the middle of an edge, we calculate:

$$\hat{e}_x(i_z) = \frac{\widehat{e}_x(i_z)}{\Delta x(i_z)} \qquad (3.80)$$

for the electric x-component and all other quantities likewise. The involved grid voltages are oriented perpendicularly to the z-axis. The value $\hat{e}_x(i_z)$ is taken to be constant along z within one cell. Then, the FV variables are obtained by:

$$d_x(i_z) = \int_{\Delta z(i_z)} \bar{\epsilon}(i_z)\hat{e}_x(i_z)\,\mathrm{d}z \Big/ \Delta z(i_z) = \bar{\epsilon}(i_z)\hat{e}_x(i_z) \qquad (3.81)$$

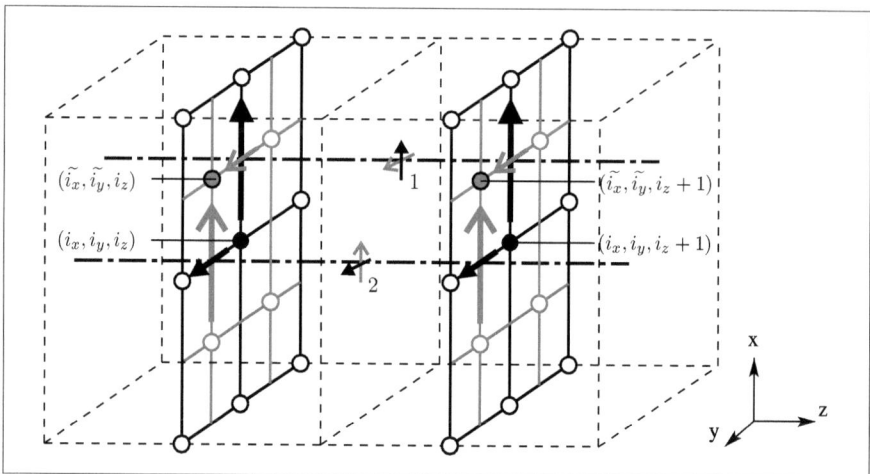

Figure 3.14: FI-FV grid setup and coupling of DoF: In the $x-y$-planes the FIT is applied. There, staggered grid doublets are employed. The x- and y-directed electric voltages of two grid points are shown in bold black arrows (full tip), the respective magnetic voltages in gray arrows (line tip). In the z-direction, the FVM is applied, which requires one grid only. The $x-y$-layers are positioned in the center of the cells in z-direction. The FVM operates along the indicated dash-dotted lines. The points 1 and 2 depict the positions where the numerical boundary fluxes are evaluated.

with all other quantities given respectively. The reconstruction of the field within a cell given by equation (3.61) can be written as:

$$E_x(i_z) = \hat{e}_x(i_z)(z - z(i_z))\, s(E_x) \tag{3.82}$$

This allows for calculating the values $E_x^\pm$, $H_y^\pm$ and $E_y^\pm$, $H_x^\pm$ at the intermediate positions 1 and 2 indicated in figure 3.14.

Matrix Formulation

The system matrix of the full hybrid FI-FV scheme can be assembled from the equations (3.17), (3.31) and (3.70). It reads:

$$\mathbf{A} = \begin{pmatrix} \mathbf{U}_H & -\mathbf{C}_E \\ \mathbf{C}_H & \mathbf{U}_E \end{pmatrix} \tag{3.83}$$

The block matrices are given by:

$$\mathbf{U}_H = \begin{pmatrix} \mathbf{U}_h & 0 & 0 \\ 0 & \mathbf{U}_h & 0 \\ 0 & 0 & 0 \end{pmatrix} \tag{3.84}$$

$$\mathbf{U}_E = \begin{pmatrix} \mathbf{U}_e & 0 & 0 \\ 0 & \mathbf{U}_e & 0 \\ 0 & 0 & 0 \end{pmatrix} \tag{3.85}$$

3.3. Finite Integration-Finite Volume Hybrid Method

$$\mathbf{C}_H = \begin{pmatrix} 0 & -(\mathbf{T}_y^h)^{-1}\mathbf{D}_\epsilon^{-1}\mathbf{P}_z^T\mathbf{T}_y^h & \mathbf{M}_{\epsilon^{-1},x}\widetilde{\mathbf{P}}_y \\ (\mathbf{T}_x^h)^{-1}\mathbf{D}_\epsilon^{-1}\mathbf{P}_z^T\mathbf{T}_x^h & 0 & -\mathbf{M}_{\epsilon^{-1},y}\widetilde{\mathbf{P}}_x \\ -\mathbf{M}_{\epsilon^{-1},x}\widetilde{\mathbf{P}}_y & \mathbf{M}_{\epsilon^{-1},y}\widetilde{\mathbf{P}}_x & 0 \end{pmatrix} \quad (3.86)$$

$$\mathbf{C}_E = \begin{pmatrix} 0 & (\mathbf{T}_y^e)^{-1}\mathbf{D}_\mu^{-1}\mathbf{P}_z\mathbf{T}_y^e & \mathbf{M}_{\mu^{-1},x}\mathbf{P}_y \\ -(\mathbf{T}_x^e)^{-1}\mathbf{D}_\mu^{-1}\mathbf{P}_z\mathbf{T}_x^e & 0 & -\mathbf{M}_{\mu^{-1},y}\mathbf{P}_x \\ -\mathbf{M}_{\mu^{-1},x}\mathbf{P}_y & \mathbf{M}_{\mu^{-1},y}\mathbf{P}_x & 0 \end{pmatrix} \quad (3.87)$$

The discrete derivative operators $\mathbf{P}_{x,y}, \widetilde{\mathbf{P}}_{x,y}$ are the respective FIT operator while $\mathbf{P}_z$ is the FV operator. The matrices $\mathbf{T}_x^e$, $\mathbf{T}_y^e$, $\mathbf{T}_x^h$ and $\mathbf{T}_y^h$ transform the FIT variables to the FV variables. They read:

$$\begin{aligned} \mathbf{T}_x^e &= \mathrm{diag}(\bar{\epsilon}(i_z)) \cdot \left[\mathrm{diag}(\Delta x(i_z))\right]^{-1}, & \mathbf{T}_y^e &= \mathrm{diag}(\bar{\epsilon}(i_z)) \cdot \left[\mathrm{diag}(\Delta y(i_z))\right]^{-1} \\ \mathbf{T}_x^h &= \mathrm{diag}(\bar{\mu}(i_z)) \cdot \left[\mathrm{diag}(\widetilde{\Delta x}(i_z))\right]^{-1}, & \mathbf{T}_y^h &= \mathrm{diag}(\bar{\mu}(i_z)) \cdot \left[\mathrm{diag}(\widetilde{\Delta y}(i_z))\right]^{-1} \end{aligned} \quad (3.88)$$

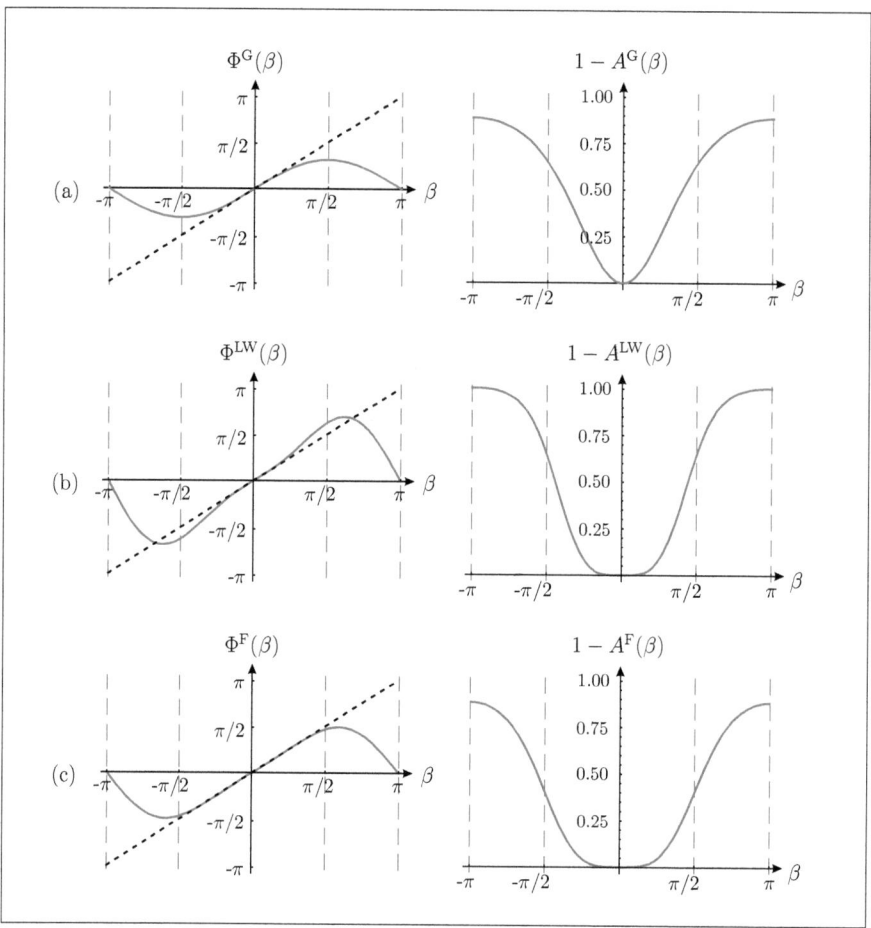

Figure 3.15: Grid dispersion and dissipation diagrams of the GODUNOV (a), LAX-WENDROFF (b) and FROMM (c) method: The plots show the numerical phase advance Φ of plane waves traveling along the z-axis. For GODUNOV's method, analogously to the FIT, all waves in the grid space lag behind the continuous case (dashed line). However, a comparison to the dispersion relation of the FIT (see fig. 3.11) reveals, that the grid resolution has to be doubled in order to yield the same result. The LAX-WENDROFF and FROMM method show a phase advance or a phase lag depending on the actual β with FROMM's method clearly performing best. For $\beta = \pm\pi$, there is no propagation which corresponds to a grid cutoff. The damping $1 - A$ increases along with β for each of the three methods. For the LAX-WENDROFF and FROMM's method, however, damping is low for practical values of β. In the long wave limit ($\beta \to 0$), the error in phase as well as the damping tend to zero.

3.4 Discontinuous Galerkin Method

For the FIT as well as for the FV methods, that have been introduced in the preceding sections, the discretization of the continuous quantities is achieved by an integration over edges, faces or volumes. The discrete state variables obtained by this procedure are assigned to the respective element in the grid space. In order to achieve a higher accuracy for a given grid, a larger number of neighboring cells has to be included in the determination of the discrete material parameters (FIT) [119] or for the reconstruction of the interface values (FVM) [38]. Using more neighbors, however, can cause problems when approaching material boundaries. Another option is the usage of a higher order approximation of the continuous quantities within each cell in the context of Finite Element Methods (FEM).

In the framework of the FEM, the infinite continuous space is projected to a space of basis functions, the Finite Element Space (FES), of a given dimensionality P. The WHITNEY Finite Element Method is well established for frequency domain problems [21]. Its application in the time-domain, however, requires the solution of a linear system of equations in every time step. This renders the method unattractive because of the associated computational costs.

The *Discontinuous* GALERKIN *Method* (DGM) avoids this drawback by choosing a local support of the basis functions. They are defined within the extent of each cell and set to zero otherwise. Generally, this leads to discontinuous approximations. The continuity across element boundaries is enforced in the weak sense via numerical fluxes. Again, the flux is not uniquely defined, but it to be determined from the values on both sides of the interface. This procedure is identical to FV methods. A comprehensive review of the DGM is found in [65].

Independently of the approximation order, only directly neighboring cells are involved in the spatial discretization procedure. The DG method, thus, always preserves a high degree of locality. Moreover, assigning different approximation orders throughout the elements of the computational domain is easily possible. This makes the method highly suitable for an hp-adaptive procedure, i.e., adapting the local grid step size as well as the local approximation order in a problem oriented fashion.

3.4.1 Semidiscrete Formulation

We seek an approximation of the space and time continuous electromagnetic field quantities $\vec{E}$ and $\vec{H}$ in the form of:

$$E_u(\vec{r},t) \approx \overline{E}_u(\vec{r},t) := \sum_{i,p} (e_u)_i^p(t)\, \varphi_i^p(\vec{r}) \tag{3.89}$$

$$H_u(\vec{r},t) \approx \overline{H}_u(\vec{r},t) := \sum_{i,p} (h_u)_i^p(t)\, \varphi_i^p(\vec{r}) \tag{3.90}$$

with $u \in \{x,y,z\}$, the approximations $\overline{E}_u$ and $\overline{H}_u$, the time-dependent coefficients $(e_u)_i^p$ and $(h_u)_i^p$ and a set of linearly independent basis functions $\{\varphi_i^p\}$ for every grid cell $\mathcal{G}_i$. The basis functions φ are required to be continuous within $\mathcal{G}_i$ and vanish otherwise:

$$\varphi_i^p(\vec{r}) := \begin{cases} \varphi^p(\vec{r}), & \vec{r} \in \mathcal{G}_i \\ 0, & \vec{r} \notin \mathcal{G}_i \end{cases} \tag{3.91}$$

Due to the cell-wise compact support of the basis functions, the approximations $\bar{E}_u$ and $\bar{H}_u$ are, in general, discontinuous in all components. Gathering the approximations for each component, results in the vectors:

$$\bar{\mathbf{E}} = (\bar{E}_x, \bar{E}_y, \bar{E}_z)^{\mathrm{T}} \quad \text{and} \quad \bar{\mathbf{H}} = (\bar{H}_x, \bar{H}_y, \bar{H}_z)^{\mathrm{T}} \tag{3.92}$$

and the cell-wise vectors of coefficients:

$$\mathbf{e}_i^p = ((e_x)_i^p, (e_y)_i^p, (e_z)_i^p)^{\mathrm{T}} \quad \text{and} \quad \mathbf{h}_i^p = ((h_x)_i^p, (h_y)_i^p, (h_z)_i^p)^{\mathrm{T}} \tag{3.93}$$

In the following, the optimality of the approximation is shortly addressed. The notations and results play an important role in section 4.4, where adaptive solution strategies for the DGM are discussed. The space spanned by the linear combinations of the basis functions is denoted by $\mathcal{V}$:

$$\mathcal{V}^P = \mathrm{span}\{\varphi^p\} \quad \text{with} \quad p = 0..P \tag{3.94}$$

The basis functions are chosen such that the approximations are elements of the space of square integrable functions $\mathcal{L}^2$. Equipping the space $\mathcal{V}$ with an inner product given by:

$$\langle u, v \rangle := \int_V u \cdot v \, \mathrm{d}V \quad \text{with} \quad u, v \in \mathcal{L}^2 \tag{3.95}$$

which induces the norm:

$$\|u\|_2 := \sqrt{\langle u, u \rangle} \tag{3.96}$$

defines a closed HILBERT space [72]. The optimal approximation to a given function in the space $\mathcal{V}$ is specified by the HILBERT projection theorem [126]:

Theorem 3.1 (HILBERT projection theorem). *Let $\mathcal{M} \subseteq \mathcal{H}$ be a closed subspace of the HILBERT space $\mathcal{H}$ and $x \in \mathcal{H}$, then there exists a unique point $m \in \mathcal{M}$ for which $\|x - m\|$ is minimized over $\mathcal{M}$. A necessary and sufficient condition for m is that the vector $(x - m)$ is orthogonal to $\mathcal{M}$.*

Thus, the optimal approximation is found via an orthogonal projection of a continuous function f to the space of basis functions $\mathcal{V}^P$. The orthogonal projection operator Π_i^p is defined by:

$$\Pi_i^p f(\vec{r}, t) := \frac{\langle \varphi_i^p, f(\vec{r}, t) \rangle}{\langle \varphi_i^p, \varphi_i^p \rangle} \varphi_i^p \tag{3.97}$$

This yields the coefficients:

$$(e_u)_i^p = \frac{\langle \varphi_i^p, E_u(\vec{r}, t) \rangle}{\langle \varphi_i^p, \varphi_i^p \rangle} \tag{3.98}$$

for the electric field and the magnetic field, respectively.

Theorem 3.2 (Optimality theorem). *The approximations given in the equations (3.89) and (3.90) with coefficients according to equation (3.98) are optimal in the sense that the approximation errors for the electric field $\mathcal{E}_u = E_u - \bar{E}_u$ and for the magnetic field $\mathcal{E}_u = H_u - \bar{H}_u$ are orthogonal to the space of basis functions $\mathcal{V}^P$:*

$$\langle \mathcal{E}_u, \varphi^p \rangle = 0 \quad \forall p \in [0, P], \varphi^p \in \mathcal{V}^P \tag{3.99}$$

3.4. Discontinuous GALERKIN Method

This completes the considerations on the optimality of the approximation.

Next, the semidiscrete weak formulation of MAXWELL's equations according to the DGM are derived [57]. The approximations (3.89) and (3.90) are substituted into FARADAY's and AMPÈRE's law in differential form. The resulting functions are tested according to the GALERKIN procedure. Using the relation $\nabla \times (\phi \vec{A}) = \phi \nabla \times \vec{A} - \vec{A} \times \nabla \phi$ and neglecting currents for the moment, they read:

$$\int_{\mathcal{G}_j} d^3\vec{r}\, \partial_t \mu \overline{\mathbf{H}} \varphi_j^q + \int_{\mathcal{G}_j} d^3\vec{r}\, \nabla \times (\varphi_j^q \overline{\mathbf{E}}) - \int_{\mathcal{G}_j} d^3\vec{r} (\nabla \varphi_j^q) \times \overline{\mathbf{E}} = 0 \quad (3.100)$$

$$\int_{\mathcal{G}_j} d^3\vec{r}\, \partial_t \epsilon \overline{\mathbf{E}} \varphi_j^q - \int_{\mathcal{G}_j} d^3\vec{r}\, \nabla \times (\varphi_j^q \overline{\mathbf{H}}) + \int_{\mathcal{G}_j} d^3\vec{r} (\nabla \varphi_j^q) \times \overline{\mathbf{H}} = 0 \quad (3.101)$$

$\forall j = 1..N$ and $\forall q = 1..P$. In order to apply STOKES' theorem to the curl-terms in the equations (3.100) and (3.101), the ambiguity of $\overline{\mathbf{E}}$ and $\overline{\mathbf{H}}$ on the cell interfaces has to be resolved. In the DG method, just like in FV methods, numerical fluxes involving unique interface values $\overline{\mathbf{E}}^*$ and $\overline{\mathbf{H}}^*$ are introduced. This allows for a representation of the curl-terms in the form of:

$$\int_{\mathcal{G}_j} d^3\vec{r}\, \nabla \times (\varphi_j^q \overline{\mathbf{H}}) = \int_{\partial \mathcal{G}_j} d^2\vec{r} (\vec{n} \times \overline{\mathbf{H}}^*) \varphi_j^q \quad (3.102)$$

$$\int_{\mathcal{G}_j} d^3\vec{r}\, \nabla \times (\varphi_j^q \overline{\mathbf{E}}) = \int_{\partial \mathcal{G}_j} d^2\vec{r} (\vec{n} \times \overline{\mathbf{E}}^*) \varphi_j^q \quad (3.103)$$

The interface values involve the state of the approximate solutions from the inside and outside of the respective interface with respect to the outward oriented normal $\vec{n}$ of $\partial \mathcal{G}_j$. This is completely analogous to the flux derivations in [77, 83] for FV methods.

The weak DG formulation of MAXWELL's equations reads:

Find $\mathbf{e}_i^p$ and $\mathbf{h}_i^p$ such that for each $j = 1..N, q = 1..P$ it holds:

$$\sum_{i,p} \delta_{ij} \left(\int_{\mathcal{G}_j} d^3\vec{r} \mu \varphi_i^p \varphi_j^q \right) d_t \mathbf{h}_i^p + \int_{\partial \mathcal{G}_j} d^2\vec{r}(\vec{n} \times \overline{\mathbf{E}}^*) \varphi_j^q - \sum_{i,p} \delta_{ij} \left(\int_{\mathcal{G}_j} d^3\vec{r} \varphi_i^p (\nabla \varphi_j^q) \right) \times \mathbf{e}_i^p = 0 \quad (3.104)$$

$$\sum_{i,p} \delta_{ij} \left(\int_{\mathcal{G}_j} d^3\vec{r} \epsilon \varphi_i^p \varphi_j^q \right) d_t \mathbf{e}_i^p - \int_{\partial \mathcal{G}_j} d^2\vec{r}(\vec{n} \times \overline{\mathbf{H}}^*) \varphi_j^q + \sum_{i,p} \delta_{ij} \left(\int_{\mathcal{G}_j} d^3\vec{r} \varphi_i^p (\nabla \varphi_j^q) \right) \times \mathbf{h}_i^p = 0 \quad (3.105)$$

---

with the KRONECKER symbol δ_{ij}.

3.4.2 Properties of the Discontinuous Galerkin Method

The equations (3.104) and (3.105) state the weak DG formulation of MAXWELL's equations in its most general form. No assumptions on the specific type of basis functions or the structure of the underlying grid were made. In the following, the DGM on Cartesian grids employing orthogonal polynomial basis functions and central fluxes is investigated.

In contrast to the Finite Volume methods, the DG method is not only characterized by the choice of the numerical flux but also by the approximation order P. The superior dispersion properties of the FVM are due to the application of upwind fluxes. Nevertheless, upwinding introduces a non-physical dissipation of energy as an undesired side-effect. Since the dispersion error of the DGM can be very efficiently reduced with the approximation order P, the central flux is employed. This yields a semidiscrete formulation that can be symplectically in time.

Basis Functions and Mass Matrix

For the given Cartesian grid, the basis functions are specified as:

$$\varphi_i^{\boldsymbol{p}}(\vec{r}) := \varphi_i^{\boldsymbol{p}}(x,y,z) = \begin{cases} \varphi^{p_x}(x)\,\varphi^{p_y}(y)\,\varphi^{p_z}(z), & (x,y,z) \in \mathcal{G}_i \\ 0, & (x,y,z) \notin \mathcal{G}_i \end{cases}, \quad \boldsymbol{p} = (p_x, p_y, p_z) \quad (3.106)$$

with the scaled Legendre polynomials:

$$\varphi^{p_x}(x) := \sqrt{2p_x+1}\,\mathcal{L}^{p_x}\left(\frac{2(x-x_0)}{\Delta x}\right) \quad (3.107)$$

of the order $p_x = 1..P_x$ (see fig. 3.16). The center of the Cartesian cell along the x-axis is denoted by x_0 and its extent by Δx. The polynomials $\varphi^{p_y}(y)$ and $\varphi^{p_z}(z)$ are defined accordingly. Using a vector notation, the basis is given by:

$$\boldsymbol{\Phi} = \begin{pmatrix} \boldsymbol{\Phi}_x \\ \boldsymbol{\Phi}_y \\ \boldsymbol{\Phi}_z \end{pmatrix} \quad \text{with} \quad \boldsymbol{\Phi}_x = (\varphi^0(x), ..., \varphi^{P_x}(x))^\mathrm{T} \quad (3.108)$$

The space $\mathcal{V}^P$ spanned by the basis functions is given by the tensor product of the respective one-dimensional spaces:

$$\mathcal{V}^P = \mathcal{V}_x^{P_x} \otimes \mathcal{V}_y^{P_y} \otimes \mathcal{V}_z^{P_z} \quad (3.109)$$

The approximation orders P_x, P_y and P_z do not have to be equal. Moreover, each of the approximation orders can be adapted separately which is an important aspect with regard to the time-adaptive refinement presented in section 4.4.

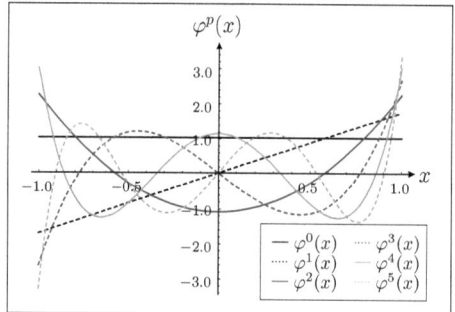

Figure 3.16: Scaled Legendre polynomials up to the order of five as basis functions for the DGM.

The scaling factor in equation (3.107) is chosen such that the basis functions are pairwise orthogonal within the interval $[-1;1]$:

$$\langle \varphi^p, \varphi^q \rangle = \int_{-1}^{1} \mathrm{d}x\,\varphi^p(x)\,\varphi^q(x) = \begin{cases} 2 & p=q, \\ 0 & p \neq q \end{cases} \quad (3.110)$$

3.4. Discontinuous GALERKIN Method

In the weak formulation (3.104) and (3.105), the terms

$$\int_{\mathcal{G}_j} \mathrm{d}^3\vec{r}\, \alpha \varphi_i^p(\vec{r}) \varphi_j^q(\vec{r}) = M_{ij,\alpha}^{pq} \quad \text{with} \quad \alpha \in \{\mu, \epsilon\} \tag{3.111}$$

$$\int_{\mathcal{G}_j} \mathrm{d}^3\vec{r}\, \varphi_i^p(\vec{r}) (\nabla \varphi_j^q(\vec{r})) = S_{ij}^{pq} \tag{3.112}$$

are readily identified as the FEM mass and stiffness terms, M and S [21]. Using an additional flux term F, the weak formulation can be expressed in matrix notation:

$$\mathbf{M}\,\mathrm{d}_t \mathbf{u} + \mathbf{F}\mathbf{u} + \mathbf{S}\mathbf{u} = 0 \tag{3.113}$$

For the classical FEM, the mass matrix $\mathbf{M}$ is sparse but its inverse is dense. Hence, the inverse is not calculated, but a linear system of equations has to be solved in every time step. The cell-wise compact support of the DGM allows for placing the KRONECKER symbol δ_{ij} in front of the mass and the stiffness terms. The corresponding matrices $\mathbf{M}$ and $\mathbf{S}$ are, therefore, guaranteed to be at most block diagonal. Substituting the tensor form (3.106) into the equation (3.111) and using the orthogonality relation (3.110), the entries of the mass matrix read:

$$M_{ij,\alpha}^{pq} = \delta_{ij}\delta_{pq}\, \alpha \Delta x_i \Delta y_i \Delta z_i \tag{3.114}$$

Due to the additional KRONECKER symbol δ_{pq}, the mass matrix becomes purely diagonal, allowing for a trivial inversion and an explicit time stepping in time domain problems[2].

System Matrix and Conservation Properties

Combining the flux and stiffness matrices $\mathbf{F}$ and $\mathbf{S}$ yields the system of ODEs of the semidiscrete DG formulation:

$$\mathbf{M}\mathrm{d}_t \mathbf{u} = \mathbf{A}\mathbf{u} \tag{3.115}$$

with the system matrix $\mathbf{A}$. As mentioned in the beginning of this section central fluxes are employed. They have been introduced in section 3.3.2 and read:

$$\vec{n} \times \vec{E}^* = \vec{n} \times \frac{1}{2}\left(\vec{E}^+ + \vec{E}^-\right), \quad \vec{n} \times \vec{H}^* = \vec{n} \times \frac{1}{2}\left(\vec{H}^+ + \vec{H}^-\right) \tag{3.116}$$

In accordance with the FIT and the central flux formulation of FVM, the equation (3.115) is rewritten in the form:

$$\mathrm{d}_t \begin{pmatrix} \mathbf{h} \\ \mathbf{e} \end{pmatrix} = \begin{pmatrix} 0 & -\mathbf{M}_\mu^{-1}\mathbf{C} \\ \mathbf{M}_\epsilon^{-1}\mathbf{C}^{\mathrm{T}} & 0 \end{pmatrix} \begin{pmatrix} \mathbf{h} \\ \mathbf{e} \end{pmatrix} \tag{3.117}$$

with the vectors of coefficients:

$$\mathbf{h} := \begin{pmatrix} \mathbf{h}_x \\ \mathbf{h}_y \\ \mathbf{h}_z \end{pmatrix} \quad \text{with} \quad \mathbf{h}_u = \left((h_u)_1^0, .., (h_u)_1^{P_1};\, ..;\, (h_u)_N^0, .., (h_u)_N^{P_N}\right)^{\mathrm{T}}, \quad u \in \{x,y,z\} \tag{3.118}$$

---

[2] The application of a local, orthogonal basis is possible only because the method allows for discontinuous solutions.

and **e** defined respectively. The local approximation order P_i is allowed to alter from cell to cell. In a full analogy to the methods presented before, the discrete curl operator mimics the structure of the continuous operator:

$$\mathbf{C} = \begin{pmatrix} 0 & -\mathbf{P}_z & \mathbf{P}_y \\ \mathbf{P}_z & 0 & -\mathbf{P}_x \\ -\mathbf{P}_y & \mathbf{P}_x & 0 \end{pmatrix} \qquad (3.119)$$

The difference to the curl operator presented for the FIT on page 33 (see [138] for details) is found in the size and structure of the discrete derivative operators $\mathbf{P}_u$. Its elements are given by:

$$\left(P_u\right)_{ij}^{pq} = \int_{\partial \mathcal{G}_j} \mathrm{d}^2 \vec{r}\, \vec{n}_u\, \varphi_i^p \varphi_j^q - \delta_{ij} \int_{\mathcal{G}_j} \mathrm{d}^3 \vec{r}\, \varphi_i^p (\nabla \varphi_j^q) \qquad (3.120)$$

The DG derivative matrix $\mathbf{P}_u$ is, thus, a $(P+1)N$ square matrix, whereas the FIT derivative matrix is a square matrix of size N. In figure 3.17 the population of the curl matrix for a computational domain of $3 \times 3 \times 3$ cells using piecewise quadratic basis functions ($P=2$) for all cells is plotted. Utilizing the properties of the tensor product basis, the Cartesian grid and the orthogonality relation (3.110), the elements of the operator for central fluxes simplify to:

$$\begin{aligned}\left(P_x\right)_{ij}^{pq} &= \delta_{p_y q_y} \delta_{p_z q_z} (A_x)_i \left[\left(\varphi_i^{p_x}(1)^* \varphi_j^{q_x}(1)\right) - \left(\varphi_i^{p_x}(-1) \varphi_j^{q_x}(-1)\right) - \delta_{ij} \int_{\mathcal{G}_{j,x}} \mathrm{d}x\, \varphi_i^{p_x}(x) \frac{\partial \varphi_j^{q_x}(x)}{\partial x} \right] \\ &= \delta_{p_y q_y} \delta_{p_z q_z} (A_x)_i \left[\frac{1}{2}(\varphi_{i+1}^{p_x}(-1) + \varphi_i^{p_x}(1)) \varphi_j^{q_x}(1) - \frac{1}{2}(\varphi_{i-1}^{p_x}(1) + \varphi_i^{p_x}(-1)) \varphi_j^{q_x}(-1) \right. \\ &\quad \left. - \delta_{ij} \int_{\mathcal{G}_{j,x}} \mathrm{d}x\, \varphi_i^{p_x}(x) \frac{\partial \varphi_j^{q_x}(x)}{\partial x} \right] \end{aligned}$$
$$(3.121)$$

with the area of the face $(A_x)_i = \Delta y_i \Delta z_i$. It can be verified that the relations:

$$\mathbf{P} = -\mathbf{P}^\mathrm{T} \quad \text{and} \quad \mathbf{C} = \mathbf{C}^\mathrm{T} \qquad (3.122)$$

hold true [57]. Applying the similarity transformation described in the equations (3.32)-(3.34), the DG system matrix can be brought to a skew-symmetric form. This allows for the same implications as for the FIT and the central flux formulation of FVM: The system of ODEs (3.117) maintains the HAMILTONIAN structure of MAXWELL's equations in continuum. The time-dependent energy $\mathcal{W}$ of the spatially discrete electromagnetic fields is defined in a full analogy to the FIT (cf. eqn. (3.27)) as:

$$\mathcal{W}(t) := \|\mathbf{u}\|_\mathrm{E} = \frac{1}{2}\left(\mathbf{h}^\mathrm{T} \mathbf{M}_\mu^{-1} \mathbf{h} + \mathbf{e}^\mathrm{T} \mathbf{M}_\epsilon^{-1} \mathbf{e}\right) \qquad (3.123)$$

Again, in the absence of currents

$$\mathrm{d}_t \mathcal{W}(t) = 0 \qquad (3.124)$$

holds true.

The DG and the FV method show a number of similarities. The approximate solution is piecewise defined and discontinuous in all components across cell boundaries. In addition, the numerical flux definitions for both methods coincide. A comparison of the equations (3.46) and (3.98), which define the degrees of freedom for each of the methods, reveals the equivalence

3.4. Discontinuous GALERKIN Method

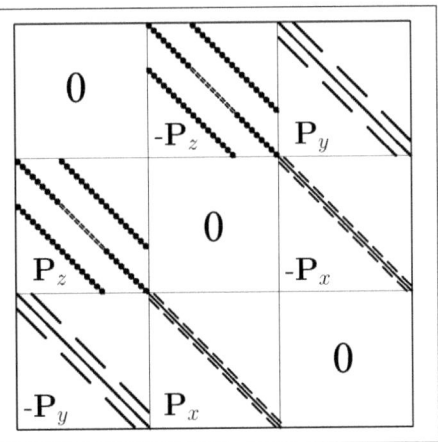

Figure 3.17: Plot of the population of the DG curl matrix (3.119) for a computational domain consisting of $3 \times 3 \times 3$ cells. The approximation order P in all cells equals two. Each derivative block $\mathbf{P}$ is square and has a size of $N(P+1)$ with elements given by the equation (3.121). The $\mathbf{P}$-matrices show a band structure assembled from blocks of size $P+1$. Due to the KRONECKER symbol δ_{ij} in (3.121), the stiffness term contributes only to the blocks on the diagonal of the respective $\mathbf{P}$-matrix while the off-diagonal blocks are determined by the flux terms only.

of the FVM employing GODUNOV slopes and the DGM employing piecewise constant basis functions.

In the particular case of particle accelerator problems, the existence of freely moving charges makes the issue of charge conservation specially relevant. In [58] it was shown that for the setup under consideration (Cartesian grid, tensor product basis and central fluxes) the DGM strictly conserves consistently defined charges in the grid space. It is concluded that pairwise commuting discrete derivative operators $\mathbf{P}_u$ are a necessary and sufficient condition for charge conservation. Since the derivative operators defined by equation (3.121) involve one spatial variable only, it is readily seen that this condition is fulfilled. Analogously to the FIT (cf. eqn. (3.23)), this allows for the definition of a discrete divergence operator $\mathbf{S}$, mimicking the continuous vector calculus identity:

$$\mathbf{SC} = \mathbf{0} \tag{3.125}$$

The DGM can be very efficiently implemented for the particular setting of a Cartesian grid, a tensor product basis and pairwise orthogonal basis functions. In addition to the simplifications arising from the diagonal mass matrix, all terms appearing in the equation (3.121) can be precalculated and stored in tables. The stiffness term can be evaluated analytically. Storing the result for each combination of p and q yields a $(P+1) \times (P+1)$-matrix. The products of the basis functions evaluated at the interval boundaries for all combinations of (p, q) and $(-1, 1)$ give rise to another four $(P+1) \times (P+1)$-matrices. Hence, the evaluation of any element of $\mathbf{P}$ reduces to five pointer operations, four summations and one multiplication. Since P does typically not exceed 5, the amount of memory necessary for the storage of the five matrices is negligible.

Dispersion Analysis

In the literature some analysis of the dispersive properties of the DGM can be found (see e.g. [3, 68, 69]). However, these consider DG formulations using upwind fluxes. Here, as presented in [55], the analysis is carried out for the non-dissipative DGM employing central fluxes. The DG curl-curl eigenvalue equation for the electric field reads:

$$\mathbf{M}_\epsilon^{-1} \mathbf{C}^T \mathbf{M}_\mu^{-1} \mathbf{C} \mathbf{e} = \omega^2 \mathbf{e} \tag{3.126}$$

As before, we use a plane wave ansatz and carry out a VON NEUMANN analysis. This allows for reducing (3.126) to a local $[3(P+1)] \times [3(P+1)]$ eigenvalue equation.

Given the approximation for the electric field within a representative cell, the approximation within the neighboring cell in positive x-direction is given by:

$$\bar{E}_{i+1}(x + \Delta x, y, z) = \bar{E}_i(x, y, z) \cdot e^{-i k_x \Delta x} \tag{3.127}$$

The approximations for all other neighbors are given accordingly. Using (3.89) and (3.106), this is rewritten as:

$$\sum_\mathbf{p} \mathbf{e}_{i+1}^\mathbf{p}(t)\, \varphi_{i+1}^{p_x}(x + \Delta x)\, \varphi_{i+1}^{p_y}(y)\, \varphi_{i+1}^{p_z}(z) = \sum_\mathbf{p} \mathbf{e}_i^\mathbf{p}(t)\, \varphi_i^{p_x}(x) \cdot e^{-i k_x \Delta x}\, \varphi_i^{p_y}(y)\, \varphi_i^{p_z}(z) \tag{3.128}$$

Hence, in (3.121), we can substitute:

$$\varphi_{i+1}^{p_x}(-1) = \varphi_i^{p_x}(-1) \cdot e^{-i k_x \Delta x} \qquad \text{and} \qquad \varphi_{i-1}^{p_x}(+1) = \varphi_i^{p_x}(+1) \cdot e^{+i k_x \Delta x} \tag{3.129}$$

Its evaluation for all (p,q) yields the local discrete derivative operator $\mathbf{P}_x$. For the approximation orders zero and one, it reads:

- $P = 0$:
$$\mathbf{P}_x = A_x \Big(\sin(k_x \Delta x) \Big) \tag{3.130}$$

- $P = 1$:
$$\mathbf{P}_x = A_x \begin{pmatrix} \sin(k_x \Delta x) & -i\sqrt{3}(\cos(k_x \Delta x) - 1) \\ i\sqrt{3}(\cos(k_x \Delta x) - 1) & -3\sin(k_x \Delta x) \end{pmatrix} \tag{3.131}$$

The operators $\mathbf{P}_y$ and $\mathbf{P}_z$ are given respectively. This is inserted into the discrete curl operator (3.119). The dispersion relations are the solutions of the resulting local eigenvalue equation. For a propagation along the coordinate u they read:

$$\Delta u \frac{\omega}{c} = \sin(k_u \Delta u) \tag{3.132}$$

$$\Delta u \frac{\omega}{c} = \begin{Bmatrix} -\sin(k_u \Delta u) - 2\sqrt{3\sin^4\left(\frac{k_u \Delta u}{2}\right) + \sin^2(k_u \Delta u)} \\ -\sin(k_u \Delta u) + 2\sqrt{3\sin^4\left(\frac{k_u \Delta u}{2}\right) + \sin^2(k_u \Delta u)} \end{Bmatrix} \tag{3.133}$$

for $P = 0, 1$. On the left hand side of the figures 3.18 and 3.19, the dispersion diagrams for approximation orders up to five are shown. A comparison of the figures 3.15a and 3.18a shows the equivalence of GODUNOV's method and the DGM employing piecewise constant basis functions regarding their dispersion properties. For orders of $P \geq 1$ unphysical solutions occur. Periodic lattices of structures that exhibit a periodicity themselves, so-called BLOCH waves appear [3]. Since an infinite, equidistant grid is assumed for the VON NEUMANN analysis, the cells form the periodic lattice, while the collocation points within the cells correspond to the periodic substructure.

In the context of simulations of MAXWELL's equations, most of the solutions are unphysical, showing a numerical phase velocity far higher than the speed of light. However, the discretization of an initial solution that is consistent with MAXWELL's equations excites only physical modes. On the right hand side of the figures 3.18 and 3.19 the phase errors of the physical modes are plotted. The error decreases rapidly with an increase of the approximation order.

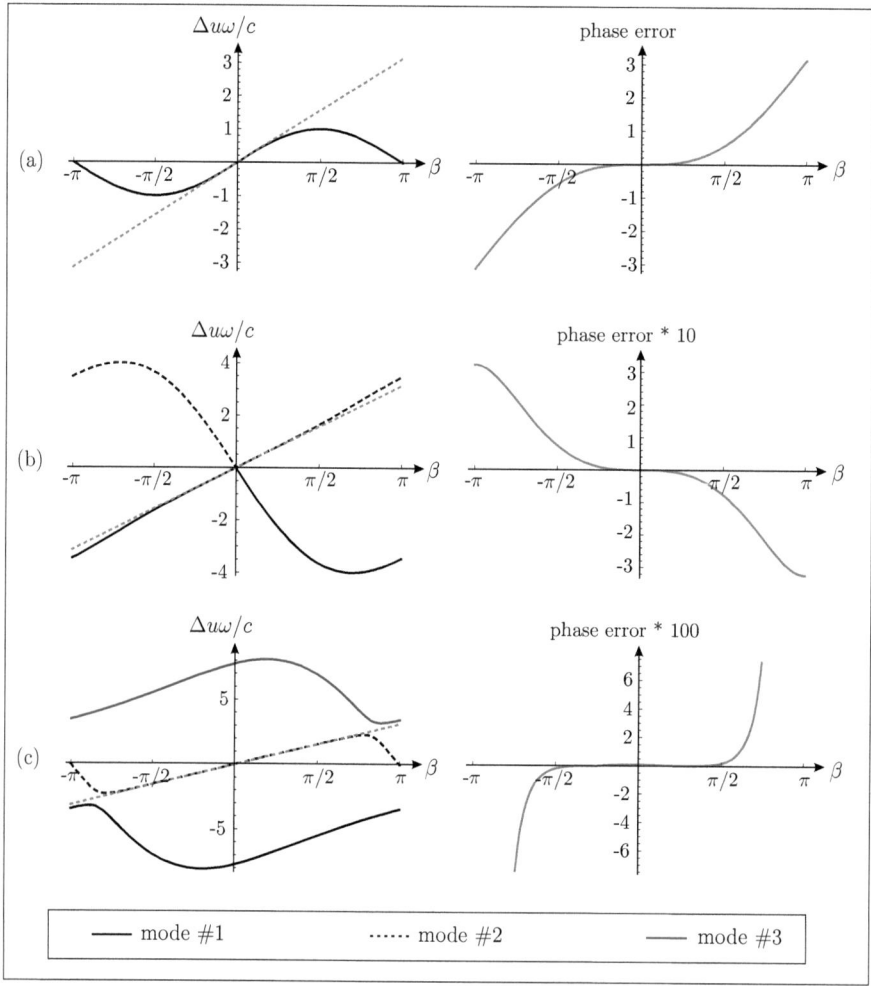

Figure 3.18: Dispersion diagrams of the DG method: The plots show the numerical phase advance and the phase error of plane waves traveling along a coordinate axis. The results correspond to (a) piecewise constant $P = 0$, (b) piecewise linear $P = 1$, and (c) piecewise quadratic $P = 2$ basis functions. Since the number of DoF per cell is equal to $P + 1$, there are $P + 1$ solutions, most of which are unphysical. For the physical solution the phase error decreases rapidly, recovering the continuous solution in the long wave limit ($\beta \to 0$).

3.4. Discontinuous GALERKIN Method

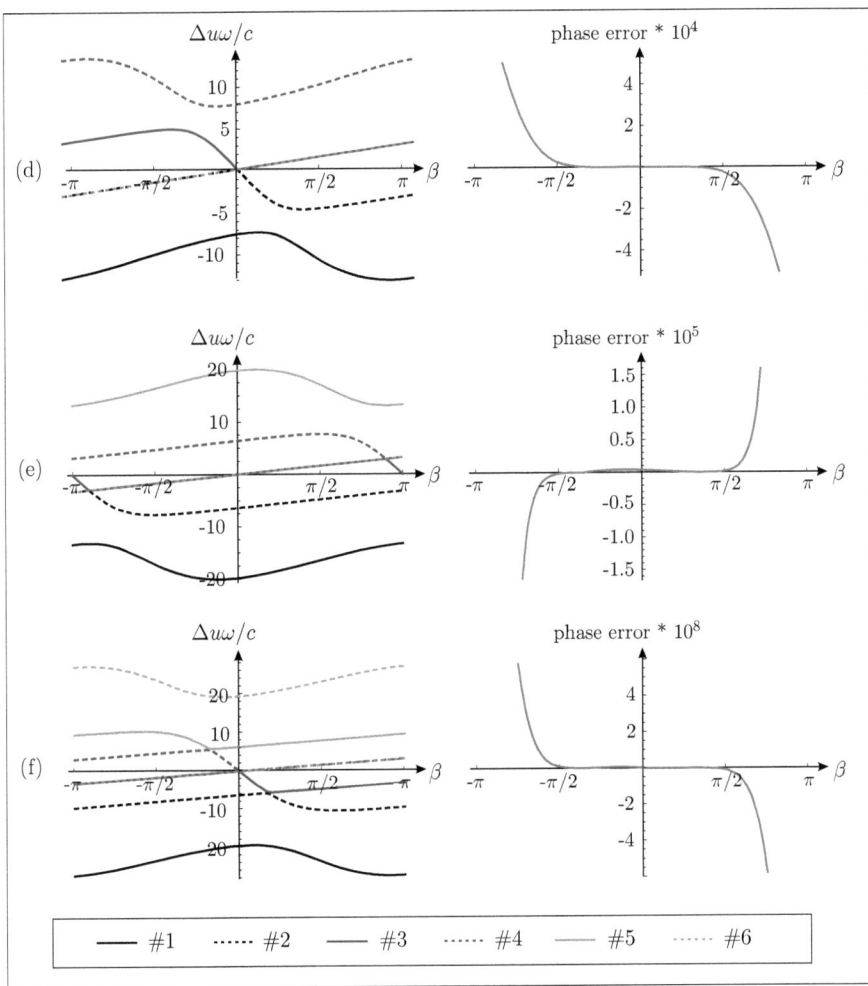

Figure 3.19: Dispersion diagrams of the DG method: The plots show the numerical phase advance and the phase error of plane waves traveling along a coordinate axis. The results correspond to basis functions of the order $P = 3$ (d), $P = 4$ (e), and $P = 5$ (f). Since the number of DoF per cell is equal to $P+1$, there are $P+1$ solutions, most of which are unphysical. For the physical solution the phase error decreases rapidly, recovering the continuous solution in the long wave limit ($\beta \to 0$).

3.5 Discretization of Time

Up to this point the spatial discretization procedures of three numerical methods have been presented. The resulting semidiscrete formulations of MAXWELL's equations for each of the methods were investigated regarding their conservation, dispersion and dissipation properties in the grid space. In this section the discretization of the temporal derivative is discussed, yielding fully discrete formulations of the MAXWELL's equations.

Discretizing the temporal variable t introduces the time instants t^n given by:

$$t^n = n \cdot \Delta t + t_0 \qquad (3.134)$$

with the time increment $\Delta t \in \mathbb{R}$ and the time step number $n \in \mathbb{N}_0$. The initial time $t_0 \in \mathbb{R}_0^+$ can be arbitrarily chosen. In the following t_0 is always set to zero. Writing the temporal evolution of the discrete electromagnetic quantities in the form:

$$\mathbf{u}^{n+1} = \mathcal{T}(\mathbf{u}^{n+1}, \mathbf{u}^n, .., \mathbf{u}^0) \qquad (3.135)$$

with $n+1$ the next step in time, allows for an iterative solution of transient phenomena. The time integration scheme described by the operator $\mathcal{T}$ is called *explicit* if $\mathcal{T}$ does not depend on the state of $\mathbf{u}$ at the following time step t^{n+1}, but only on the states at the instants t^n and before. Otherwise it is called *implicit*. Due to their minor computational costs, only explicit schemes are considered in this work.

In the preceding sections semidiscrete systems of ODEs in matrix form were derived. Explicitly writing down the dependence on the temporal variable t, they read:

$$\mathrm{d}_t \mathbf{u}(t) = \mathbf{A}\mathbf{u}(t) \qquad (3.136)$$

Next, a discretization of the temporal variable t according to (3.134) and a replacement of the time derivative by a difference approximation is carried out for each of the semidiscrete systems (3.31), (3.67) and (3.115). In the sections 3.2–3.4 it was emphasized that the possibility of a symplectic time integration depends on the eigenvalue distribution of the system matrix $\mathbf{A}$. Therefore, the FIT and the central flux formulation of the DGM, both having system matrices with pairs of purely imaginary eigenvalues, are discussed together. The upwind flux formulation of the FI-FV scheme, exhibiting mixed complex eigenvalues of the system matrix, is treated separately.

3.5.1 Time Integration of Hamiltonian Systems

For the FIT as well as for the DGM, the conservation of the semidiscrete energy $\mathcal{W}$ was stated in the equations (3.30) and (3.124). For the numerical integration of HAMILTONIAN systems the so-called *symplectic integrators* are widely used. Symplectic time integration schemes conserve the occupied volume in the phase space $\mathbb{S}$ spanned by the electric field $\vec{E}$ and the magnetic field $\vec{H}$. This in turn guarantees for no secular increase of the error in the total energy. See [145, 146] for details.

The simplest symplectic time integration scheme is the *leap-frog* scheme. It is commonly applied for the solution of transient problems with the FIT (see e.g. [139]). Recently, the leap-frog

3.5. Discretization of Time

scheme was also proposed as one possible method for performing symplectic time integration in conjunction with the central flux DGM [57].

The leap-frog scheme makes use of a central difference approximation of the derivative in time. The fully discrete system of equations reads:

$$\frac{(\mathbf{u}')^{n+1} - (\mathbf{u}')^n}{\Delta t} = \mathbf{A}'(\mathbf{u}')^n \quad (3.137)$$

with

$$(\mathbf{u}')^n = \begin{pmatrix} (\mathbf{h}')^n \\ (\mathbf{e}')^{n+1/2} \end{pmatrix} \quad (3.138)$$

Here, the skew-symmetric system matrix $\mathbf{A}'$ was used to emphasize the HAMILTONIAN structure of the semidiscrete FIT and DG formulation of MAXWELL's equations. In the following, the primes are dropped. The vectors $\mathbf{h}^n$ and $\mathbf{e}^{n+1/2}$ represent the DoF of either the FIT or the DGM at the respective full or half time step[3]. Arranging the equation (3.137) in the form of (3.135) yields the update equation:

$$\mathbf{u}^{n+1} = \mathbf{T}\mathbf{u}^n \quad (3.139)$$

with the *iteration matrix* $\mathbf{T}$:

$$\mathbf{T} = \begin{pmatrix} \mathbf{I} & -\Delta t \mathbf{M}_{\mu^{-1}}^{1/2} \mathbf{C} \mathbf{M}_\epsilon^{-1/2} \\ \Delta t \mathbf{M}_\epsilon^{-1/2} \mathbf{C}^T \mathbf{M}_{\mu^{-1}}^{1/2} & \mathbf{I} - \Delta t^2 \mathbf{M}_\epsilon^{-1/2} \mathbf{C}^T \mathbf{M}_{\mu^{-1}} \mathbf{C} \mathbf{M}_\epsilon^{-1/2} \end{pmatrix} \quad (3.140)$$

where $\mathbf{C}$ and $\mathbf{M}$ are the curl and material/mass matrices of the FIT or DGM respectively. The symplectic property of phase space volume conservation can be expressed by the condition:

$$\mathbf{T}^T \begin{pmatrix} \mathbf{0} & \mathbf{I} \\ -\mathbf{I} & \mathbf{0} \end{pmatrix} \mathbf{T} \stackrel{!}{=} \begin{pmatrix} \mathbf{0} & \mathbf{I} \\ -\mathbf{I} & \mathbf{0} \end{pmatrix} \quad (3.141)$$

which holds true for the iteration matrix $\mathbf{T}$ given above.

The leap-frog scheme preserves the charge conservation properties of the FIT and the central flux DGM. For each of the methods a consistent discrete divergence operator $\mathbf{S}$ can be defined. The left-application of this operator to the update equation for the magnetic flux $\mathbf{b}$, yields [33]:

$$\mathbf{S}\mathbf{b}^{n+1} = \mathbf{S}\mathbf{b}^n - \Delta t \mathbf{S}\mathbf{C}\mathbf{e}^{n+1/2} \quad (3.142)$$

Due to the identity $\mathbf{SC} = \mathbf{0}$ (cf. eqns. (3.23), (3.125)) no magnetic charges appear. The conservation of electric charges is verified in the same manner.

The iteration matrix $\mathbf{T}$ is determined by the system matrix $\mathbf{A}$ of the respective spatial discretization method and the chosen difference approximation to the discretization of the time derivative. Comparing the eigenvalues of the iteration matrix λ_T with those of the system matrix λ_A allows for establishing the relation:

$$\lambda_T = \alpha \pm \sqrt{\alpha^2 - 1} \quad \text{with} \quad \alpha = 1 + \frac{1}{2}(\Delta t \lambda_A)^2 \quad (3.143)$$

If the stability condition is fulfilled[4], the leap-frog scheme maps the pairs of purely imaginary eigenvalues of the FIT and DG system matrix to the unit circle in the complex plane (see fig. 3.20). It is emphasized that all considerations assume lossless materials.

---
[3] In the update equation for the magnetic field $\mathbf{h}$ the central difference in time makes use of the time stages $n+1$ and n, *leapfrogging* the time level $n+1/2$ of the electric field.
[4] The stability of numerical approximation methods is addressed in the following section.

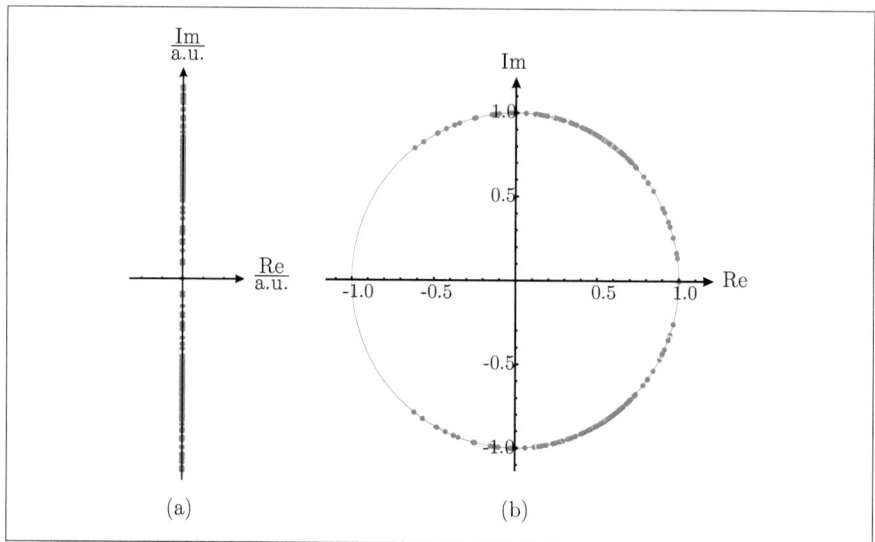

Figure 3.20: Exemplary eigenvalue distribution of the FIT and DG system matrix (a) and iteration matrix (b) in the complex plane: The leap-frog scheme maps each pair of eigenvalues of the system matrix λ_A to the unit circle in the complex plane.

Applying the VON NEUMANN analysis to the central difference approximation comprised by the leap-frog scheme yields the dispersion relation of the temporal discretization:

$$\sum_u \left[(k_u)^2\right] = \frac{1}{c^2}\left(\frac{\sin\left(\frac{\omega \Delta t}{2}\right)}{\frac{\Delta t}{2}}\right)^2 \quad \text{with} \quad u \in \{x, y, z\} \quad (3.144)$$

For the FIT and the DGM employing piecewise constant basis functions the dispersion relations of the fully discretized equations read:

- FIT:

$$\sum_u \left[\frac{\left(\sin\left(\frac{k_u \Delta u}{2}\right)\right)^2}{\left(\frac{\Delta u}{2}\right)^2}\right] = \frac{1}{c^2}\left(\frac{\sin\left(\frac{\omega \Delta t}{2}\right)}{\frac{\Delta t}{2}}\right)^2 \quad (3.145)$$

- DG $P = 0$:

$$\sum_u \left[\frac{(\sin(k_u \Delta u))^2}{\Delta u^2}\right] = \frac{1}{c^2}\left(\frac{\sin\left(\frac{\omega \Delta t}{2}\right)}{\frac{\Delta t}{2}}\right)^2 \quad (3.146)$$

with $u \in \{x, y, z\}$. The relations are equal except for a division by two of the spatial step size as a result of the staggered grid setup of the FIT.

3.5.2 Time Integration of Non-Hamiltonian Systems

In section 3.3.2 it was shown that the system matrix of the hybrid Finite Integration-Finite Volume method employing upwind fluxes given in (3.83) on page 44 does not expose the structure of a HAMILTONIAN system. Its eigenvalues are general complex numbers.

Leap-Frog Time Integrator

A symplectic time integration method cannot be applied to a system with general complex eigenvalues, which is demonstrated in the following. In figure 3.21a the eigenvalues of the FI-FV system matrix λ_A are plotted in the complex plane. If the leap-frog scheme is applied to the time integration, the relation between the eigenvalues of the system matrix and the iteration matrix is given by equation (3.143). The two parts of the relation map the pairs of eigenvalues once to the inside $(\alpha - \sqrt{\alpha^2 - 1})$ of the unit circle and once to the outside $(\alpha + \sqrt{\alpha^2 - 1})$. The eigenvalue distribution of the iteration matrix is shown in figure 3.21b. A discussion of stability is conducted in section 3.6. Nevertheless, it is evident from equation (3.139) that eigenvalues with $|\lambda_T| > 1$ lead to an exponential amplification of the respective mode. Consequently, the fully discrete method is unstable for any choice of the time step Δt.

Runge-Kutta Time Integrator

The RUNGE-KUTTA-type methods (RK) [61] are possible choices for the time-integration of the FI-FV scheme. The update equations for the *predictor-corrector* method (RK2) and the classical RK4 read:

$$\text{RK2:} \quad \mathbf{u}^{n+1} = \left(\mathbf{I} + \Delta t \mathbf{A} + \frac{(\Delta t \mathbf{A})^2}{2} \right) \mathbf{u}^n \tag{3.147}$$

$$\text{RK4:} \quad \mathbf{u}^{n+1} = \left(\mathbf{I} + \Delta t \mathbf{A} + \frac{(\Delta t \mathbf{A})^2}{2} + \frac{(\Delta t \mathbf{A})^3}{6} + \frac{(\Delta t \mathbf{A})^4}{24} \right) \mathbf{u}^n \tag{3.148}$$

In the figures 3.21c and 3.21d the eigenvalue distribution of the iteration matrices for both methods are shown. The relations of the eigenvalues of the system matrix λ_A and those of the iteration matrix λ_T read:

$$\lambda_T = 1 + \Delta t \lambda_A + \frac{(\Delta t \lambda_A)^2}{2} \left[+ \frac{(\Delta t \lambda_A)^3}{6} + \frac{(\Delta t \lambda_A)^4}{24} \right] \tag{3.149}$$

3.5.3 Time Integration Applying Split Operator Techniques

Split Operator Techniques

The exact solution of the semidiscrete equation (3.136) within one time step Δt is formally given by:

$$\mathbf{u}^{n+1} = e^{-\Delta t \mathbf{A}} \mathbf{u}^n \tag{3.150}$$

Performing a splitting of the matrix operator $\mathbf{A}$ such that $\mathbf{A} = \mathbf{A}_1 + \mathbf{A}_2$ this solution reads:

$$\mathbf{u}^{n+1} = e^{-\Delta t (\mathbf{A}_1 + \mathbf{A}_2)} \mathbf{u}^n \tag{3.151}$$

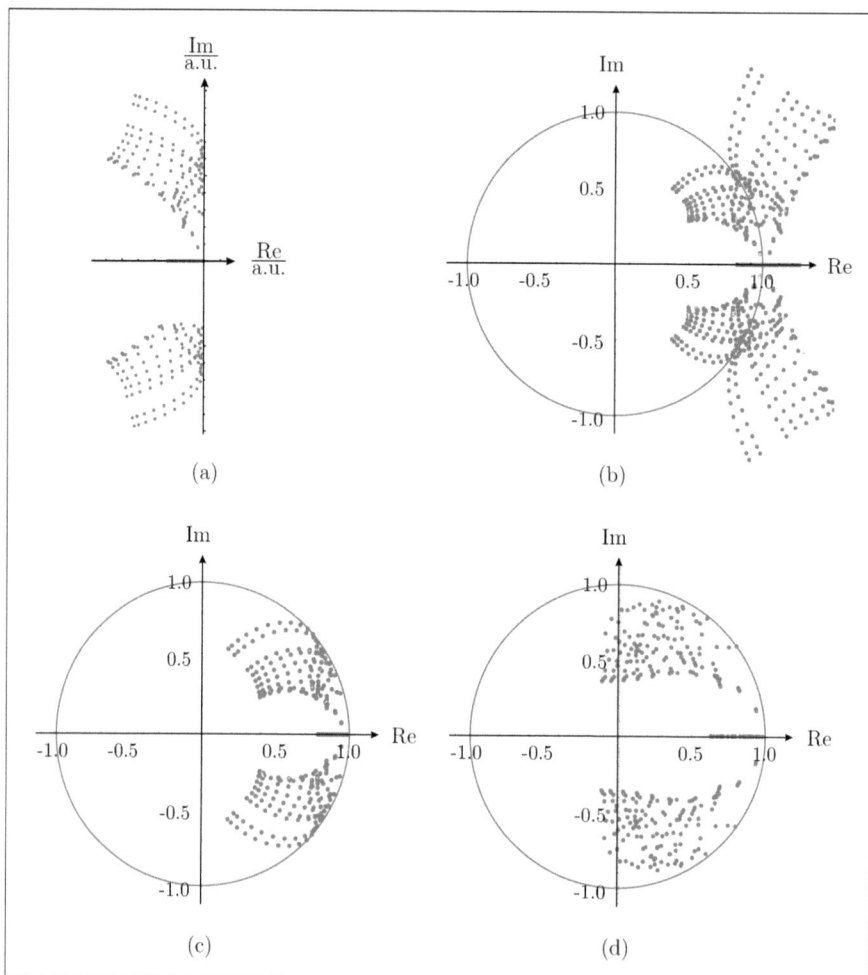

Figure 3.21: Exemplary eigenvalue distribution of the FI-FV hybrid system matrix (a) and iteration matrices (b)-(d) in the complex plane: The iteration matrices correspond to the application of the leap-frog scheme (b), the predictor-corrector scheme (c) and the classical Runge-Kutta-4 method (d). Only the latter two methods lead to a stable time stepping scheme.

In section 3.3, the development of the hybrid FI-FV method was motivated from a directional separation of the continuous curl operator in equation (3.42) on page 37. In the following, the matrices $\mathbf{A}_1$ and $\mathbf{A}_2$ are defined as the transverse and longitudinal operators $\mathbf{A}_\mathrm{T}$ and $\mathbf{A}_\mathrm{L}$. This leads to a *longitudinal-transverse splitting* (LT) of the system matrix [80]. The separation of

3.5. Discretization of Time

the update operator according to:

$$e^{-\Delta t \mathbf{A}} = e^{-\Delta t (\mathbf{A}_T + \mathbf{A}_L)} = e^{-\Delta t \mathbf{A}_T} \cdot e^{-\Delta t \mathbf{A}_L} + \mathcal{O}(\Delta t^2) \qquad (3.152)$$

leads to the first order accurate GODUNOV splitting [59]. A second order accurate time update scheme results from the multiplicative STRANG splitting procedure [121]:

$$e^{-\Delta t \mathbf{A}} = e^{-\Delta t/2 \mathbf{A}_T} \cdot e^{-\Delta t \mathbf{A}_L} \cdot e^{-\Delta t/2 \mathbf{A}_T} + \mathcal{O}(\Delta t^3) \qquad (3.153)$$

The construction of higher order split operator methods is addressed in [57, 146].

Substituting the equations (3.152), (3.153) into (3.150) and introducing the transverse and longitudinal iteration matrices $\mathbf{T}_T(\Delta t)$ and $\mathbf{T}_L(\Delta t)$ yields:

$$\mathbf{u}^{n+1} = \mathbf{T}_T(\Delta t)\, \mathbf{T}_L(\Delta t)\, \mathbf{u}^n \qquad (3.154)$$

$$\mathbf{u}^{n+1} = \mathbf{T}_T(\Delta t/2)\, \mathbf{T}_L(\Delta t)\, \mathbf{T}_T(\Delta t/2)\, \mathbf{u}^n \qquad (3.155)$$

where the error terms were dropped. The additional computational costs of the STRANG splitting, arising from the separation of the transverse update, can be minimized by merging the last and first half update of consecutive time steps:

$$\begin{aligned}
e^{-2\Delta t \mathbf{A}} &= e^{-\Delta t/2 \mathbf{A}_T} \cdot e^{-\Delta t \mathbf{A}_L} \cdot \underbrace{e^{-\Delta t/2 \mathbf{A}_T} \cdot e^{-\Delta t/2 \mathbf{A}_T}}_{e^{-\Delta t \mathbf{A}_T}} \cdot e^{-\Delta t \mathbf{A}_L} \cdot e^{-\Delta t/2 \mathbf{A}_T} + \mathcal{O}(\Delta t^3) \\
&= e^{-\Delta t/2 \mathbf{A}_T} \cdot e^{-\Delta t \mathbf{A}_L} \cdot e^{-\Delta t \mathbf{A}_T} \cdot e^{-\Delta t \mathbf{A}_L} \cdot e^{-\Delta t/2 \mathbf{A}_T} + \mathcal{O}(\Delta t^3)
\end{aligned} \qquad (3.156)$$

Time Integration of the FI-FV Scheme

In the update equations (3.154) and (3.155) the two-dimensional transverse problem is separated from the one-dimensional longitudinal problem. This allows for applying different time integration methods to each of the subproblems. In figure 3.22, the eigenvalues of $\mathbf{A}$, $\mathbf{A}_T$ and $\mathbf{A}_L$ are displayed. Since the transverse problem is discretized using the FIT, the eigenvalues of the resulting transverse system matrix $\mathbf{A}_T$ are pairwise purely imaginary, whereas the eigenvalues of the longitudinal problem are pairwise conjugate complex. The transverse problem can be symplectically integrated in time using the leap-frog scheme. In contrast to the application of a RUNGE-KUTTA time integrator to the full three-dimensional system, this limits numerical dissipation effects to the longitudinal part. The iteration matrix $\mathbf{T}_T$ is similar to the iteration matrix $\mathbf{T}$ given in equation (3.140), except for $\mathbf{C}$ being replaced by a transverse curl matrix $\mathbf{C}_T$.

For the time integration of the longitudinal problem, the LAX-WENDROFF method is applied. The time discrete update equation reads:

$$\vec{u}(k)^{n+1} = \vec{u}(k)^n + \frac{\Delta t}{\Delta z(k)} \left[f(k+1/2)^{n+1/2} - f(k-1/2)^{n+1/2} \right], \quad \forall k = [1..N_z] \qquad (3.157)$$

The LAX-WENDROFF method combines an explicit EULER step in time with numerical fluxes evaluated at the time level $n + 1/2$. This requires an extrapolation in time for the calculation of the boundary values given by:

$$\left(E_x(k)^{n+1/2}\right)^\pm = \frac{d_x(k)^n}{\epsilon} \mp \frac{\Delta z}{2} \partial_z E_x + \frac{\Delta t}{2} \partial_t E_x \qquad (3.158)$$

for E_x and all other quantities likewise. The temporal derivative can be cast into a spatial derivative using the underlying PDE. This is referred to as the CAUCHY-KOWALEWSKI procedure [77]. It yields:

$$\left(E_x(k)^{n+1/2}\right)^{\pm} = \frac{d_x(k)^n}{\epsilon} \mp \frac{\Delta z}{2}\partial_z E_x - \frac{\Delta t}{2}\frac{1}{\epsilon}\partial_z H_y \qquad (3.159)$$

where the temporal derivate of E_x was replaced by the spatial derivative of H_y according to AMPÈRE's law (2.6). The spatial derivative is replaced by the slope approximation specified in the equations (3.63), (3.64) and (3.65) for the GODUNOV, LAX-WENDROFF and FROMM method:

$$\left(E_x(k)^{n+1/2}\right)^{\pm} = \frac{d_x(k)^n}{\epsilon} \mp \frac{\Delta z}{2}s(E_x) - \frac{\Delta t}{2}\frac{1}{\epsilon}s(H_y) \qquad (3.160)$$

The update (3.157) has to be carried out for all z-directed lines given by the combinations of (i_x, i_y) with $i_x = [1..N_x]$ and $i_y = [1..Ny]$ (see fig. 3.14). The iteration matrix $\mathbf{T}_\mathrm{L}$ reads:

$$\mathbf{T}_\mathrm{L}(\Delta t) = \mathbf{I} + \Delta t \mathbf{A}_\mathrm{L} = \begin{pmatrix} \mathbf{I} + \Delta t (\mathbf{U}_H)_\mathrm{L} & -\Delta t\, (\mathbf{C}_E)_\mathrm{L} \\ \Delta t\, (\mathbf{C}_H)_\mathrm{L} & \mathbf{I} + \Delta t (\mathbf{U}_E)_\mathrm{L} \end{pmatrix} \qquad (3.161)$$

with the respective longitudinal operators $(\mathbf{U}_{H/E})_\mathrm{L}$, $\left(\mathbf{C}_{H/E}\right)_\mathrm{L}$. The relation between the eigenvalues of $\mathbf{A}_\mathrm{L}$ and $\mathbf{T}_\mathrm{L}$ reads:

$$\lambda_\mathrm{T} = 1 + \Delta t \lambda_\mathrm{A} \qquad (3.162)$$

Both eigenvalue distributions are plotted in figure 3.23.

A VON NEUMANN analysis yields the phase and amplification functionss:

$$\begin{aligned}\Phi(\omega \Delta t) &= \sin(\omega \Delta t) & (3.163)\\ A(\omega \Delta t) &= \exp(\cos(\omega \Delta t) - 1) & (3.164)\end{aligned}$$

Combining these with the spatial phase and amplification properties of the FVMs given in the equations (3.74)-(3.79) on page 43, the properties of the fully discretized one-dimensional subproblem emerge.

3.5. Discretization of Time

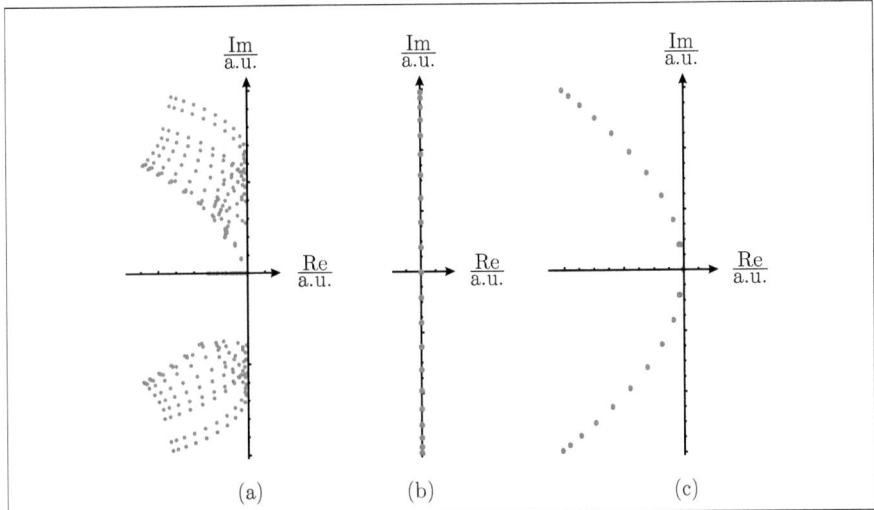

Figure 3.22: Exemplary eigenvalue distribution of the FI-FV hybrid system matrix $\mathbf{A}$ (a) and of the transverse and longitudinal matrices $\mathbf{A}_T$ (b) and $\mathbf{A}_L$ (c) in the complex plane.

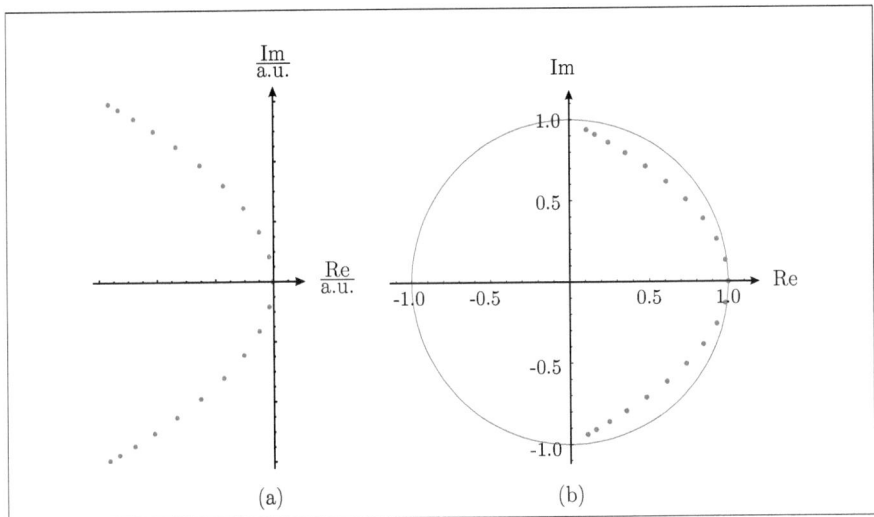

Figure 3.23: Exemplary eigenvalue distribution of the longitudinal system matrix $\mathbf{A}_L$ (a) and the iteration matrix $\mathbf{T}_L$ for an application of the explicit EULER scheme (b).

3.6 Consistency, Stability and Convergence

Up to this point, three (classes of) methods for the discretization of MAXWELL's equations were introduced. In this section the *consistency, stability* and *convergence* of these methods are addressed.

Colloquially, an approximate equation is called consistent if it equals the true equation in the limit of $\Delta u \to 0$ and $\Delta t \to 0$. The solution to an approximation equation is called convergent if it equals the solution to the true equation in the limit of $\Delta u \to 0$ and $\Delta t \to 0$ ([77] p. 189). Note that consistency refers to the approximate equation but convergence refers to its solution. Since the latter depends on the initial conditions and on the time stepping, they are not identical. If the discrete energy grows indefinitely during the time stepping the method is unstable and though the approximate equation might be consistent, no convergence can be attained.

For a definition of these properties, the *truncation error* $\boldsymbol{\mathcal{T}}$ is introduced as:

$$\boldsymbol{\mathcal{T}}^{n+1} := \hat{\mathbf{u}}^{n+1} - \mathbf{T}\hat{\mathbf{u}}^n \qquad (3.165)$$

with the exact solution $\hat{\mathbf{u}}$ at the time levels n and $n+1$.

Definition 3.1 (Consistency). A scheme of the form (3.139) is consistent if and only if for the truncation error $\boldsymbol{\mathcal{T}}$ defined in equation (3.165) it holds true:

$$\|\boldsymbol{\mathcal{T}}^{n+1}\| \xrightarrow[\Delta u, \Delta t \to 0]{} 0 \qquad (3.166)$$

for some norm $\|\cdot\|$.

If k and l are the largest integers which fulfill

$$\|\boldsymbol{\mathcal{T}}^{n+1}\| \leq r\left[(\Delta t)^k + (\Delta u)^l\right] \qquad \text{for} \qquad \Delta u, \Delta t \to 0,\ r \in \mathbb{R}^+ \qquad (3.167)$$

the respective scheme is said to be k-th order accurate in time and l-th order accurate in space.

While consistency refers to the error committed within a single time step for exact initial data, stability is related to the accumulation of errors. Defining the error as:

$$\boldsymbol{\mathcal{E}}^n := \hat{\mathbf{u}}^n - \mathbf{u}^n \qquad (3.168)$$

the error at time level $n+1$ reads:

$$\begin{aligned}\boldsymbol{\mathcal{E}}^{n+1} &= \|\hat{\mathbf{u}}^{n+1} - \mathbf{u}^{n+1}\| = \|\hat{\mathbf{u}}^{n+1} - \mathbf{T}\hat{\mathbf{u}}^n + \mathbf{T}\hat{\mathbf{u}}^n - \mathbf{u}^{n+1}\| \\ &= \|\hat{\mathbf{u}}^{n+1} - \mathbf{T}\hat{\mathbf{u}}^n + \mathbf{T}\hat{\mathbf{u}}^n - \mathbf{T}\mathbf{u}^n\| \\ &\leq \|\boldsymbol{\mathcal{T}}^{n+1}\| + \|\mathbf{T}\hat{\mathbf{u}}^n - \mathbf{T}\mathbf{u}^n\| = \|\boldsymbol{\mathcal{T}}^{n+1}\| + \|\mathbf{T}\boldsymbol{\mathcal{E}}^n\|\end{aligned} \qquad (3.169)$$

The error at this time level is composed of two parts. Namely, the truncation error committed in the current time step and the accumulated error. Hence, for the definition of stability it follows:

Definition 3.2 (Stability). A scheme of the form (3.139) is stable if and only if for the error defined in (3.168) it holds true:

$$\|\mathbf{T}\boldsymbol{\mathcal{E}}^n\| \leq \|\boldsymbol{\mathcal{E}}^n\| \qquad (3.170)$$

Finally, the definition of convergence reads:

Definition 3.3 (Convergence). A scheme of the form (3.139) provides a convergent approximation if and only if for the error defined in (3.168) it holds:

$$\|\boldsymbol{\mathcal{E}}^n\| \xrightarrow[\Delta u, \Delta t \to 0]{} 0 \qquad \text{for any} \quad n\Delta t = t \in (t_0, T] \qquad (3.171)$$

3.6. Consistency, Stability and Convergence

Consistency

The consistency of a spatial discretization method can be shown by an investigation of the grid dispersion relation. The semidiscrete dispersion relations for the FIT and the FVMs are given in the equation (3.39) and equations (3.74), (3.76) and (3.78). They are also plotted in the figures 3.11 and 3.15. For the DGM, the dispersion relations for $P \leq 1$ are given in (3.132) and (3.133). For $P \leq 5$, they are illustrated in the figures 3.18 and 3.19. A consistent spatial discretization method has to recover the true equation and thus the physical phase speed in the limit of $\Delta u \to 0$. This is clearly fulfilled for all presented methods. Using (3.40), the error in the numerical phase advance of the FIT for waves propagating along a coordinate reads:

$$\Delta \beta = k_u \Delta u - 2 \sin\left(\frac{k_u \Delta u}{2}\right) \tag{3.172}$$

It can be expanded into a power series of the spatial step size Δu about the point $\beta = 0$:

$$\Delta \beta \approx \frac{k_u^3 \Delta u^3}{24} - \frac{k_u^5 \Delta u^5}{1920} \tag{3.173}$$

The leading term of this long wave approximation, provides a natural way of defining the order of the dispersion error [38]. For the three FVMs this procedure yields:

- GODUNOV:
$$\Delta \beta \approx \frac{k_u^3 \Delta u^3}{6} - \frac{k_u^5 \Delta u^5}{120} \tag{3.174}$$

- LAX-WENDROFF:
$$\Delta \beta \approx \frac{k_u^3 \Delta u^3}{6} - \frac{k_u^5 \Delta u^5}{120} \tag{3.175}$$

- FROMM:
$$\Delta \beta \approx \frac{k_u^3 \Delta u^3}{12} - \frac{13 k_u^5 \Delta u^5}{240} \tag{3.176}$$

The dominant error term is of the order three. All schemes are, thus, second order accurate in the long wave limit ($\beta \to 0$) regarding their dispersion properties. See figure 3.24 for an illustration.

A similar expansion can be done for the dissipation error of the FVMs (eqns. (3.75), (3.77) and (3.79)). It reveals that GODUNOV's method is first order accurate and the LAX-WENDROFF and FROMM scheme are third order accurate regarding their dissipation properties. This makes clear that the order of accuracy is not uniquely determined for a numerical method but it depends on the property under consideration and possibly the norm applied to study it.

The power expansion of the DGM dispersion error for waves propagating along a coordinate for the orders P from zero to five yields:

$$\Delta \beta_0 \approx \frac{k_u^3 \Delta u^3}{6} - \frac{k_u^5 \Delta u^5}{120} \qquad \Delta \beta_1 \approx \frac{k_u^3 \Delta u^3}{48} - \frac{7 k_u^5 \Delta u^5}{15360} \tag{3.177}$$

$$\Delta \beta_2 \approx \frac{k_u^7 \Delta u^7}{f_5} + \frac{k_u^9 \Delta u^9}{f_6} \qquad \Delta \beta_3 \approx \frac{k_u^7 \Delta u^7}{f_7} + \frac{k_u^9 \Delta u^9}{f_8} \tag{3.178}$$

$$\Delta \beta_4 \approx \frac{k_u^{11} \Delta u^{11}}{f_9} + \frac{k_u^{13} \Delta u^{13}}{f_{10}} \qquad \Delta \beta_5 \approx \frac{k_u^{11} \Delta u^{11}}{f_{11}} + \frac{k_u^{13} \Delta u^{13}}{f_{12}} \tag{3.179}$$

The value of f strongly increases with the index [55]. The order of accuracy does not change continuously but only with every other increment of P. This allows for establishing the empirical order of accuracy of the central flux DGM in dependence of P:

$$\frac{\Delta\beta}{\Delta u} \sim \mathcal{O}(\Delta u^{2P+M}) \quad \text{with} \quad M = 2 \cdot \text{mod}\left(\frac{P+1}{2}\right) \tag{3.180}$$

This is illustrated in figure 3.25. A similar expression for the DGM employing upwind fluxes was found in [9, 68].

For the time integration schemes, consistency follows from their phase and amplification functions given in the equation (3.144) for the leap-frog scheme, and in the equations (3.163) and (3.164) for the explicit EULER method. The dispersion relation of the leap-frog scheme in time resembles that of the FIT in space. The relations of the explicit EULER method resemble those of GODUNOV's method. Following the argument above, both schemes are consistent, being second order accurate regarding their dispersion properties; the explicit EULER method, moreover, being first order accurate concerning the dissipation properties.

Stability

In general, explicit time integration schemes are not unconditionally stable. Their stability is restricted by an upper boundary of the time step Δt. For the leap-frog scheme this limit can be derived from the dispersion relations (3.145) and (3.146). Recalling that we assumed time-harmonic fields of the form:

$$\mathbf{u}(t) \sim e^{i\omega t} \tag{3.181}$$

the angular frequency ω has to be real-valued in order to avoid an exponential growth of the discrete field quantities. Thus, we require that:

$$0 \leq \sin^2\left(\frac{\omega \Delta t}{2}\right) \leq 1 \tag{3.182}$$

Substituting the spatial terms in (3.145), (3.146) with their upper limits:

$$\sin\left(\frac{k_u \Delta u}{2}\right) \equiv 1 \quad \text{and} \quad \sin(k_u \Delta u) \equiv 1 \tag{3.183}$$

3.6. Consistency, Stability and Convergence

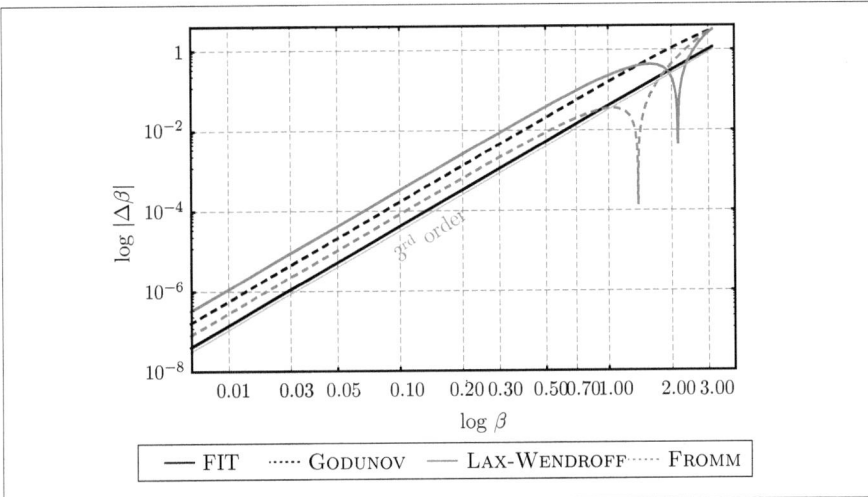

Figure 3.24: Dispersion error of the FIT and FVMs: In the long wave limit of $\beta \to 0$ the dispersion error of all methods tends to zero in third order.

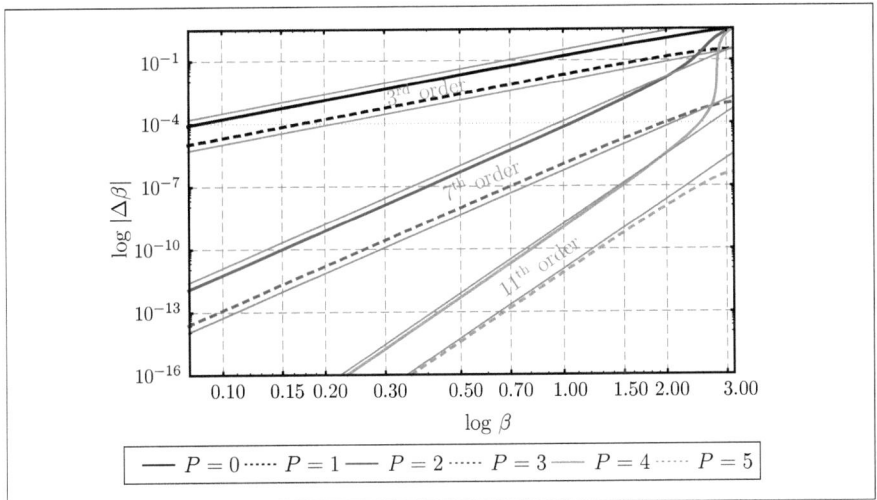

Figure 3.25: Dispersion error of the DGM: The order of accuracy of the dispersion error depends on the approximation order P and is given in the equation (3.180). For $P = 0, 1$ the error tends to zero in third order, for $P = 2, 3$ in seventh order, and for $P = 4, 5$ in eleventh order.

yields the maximally stable time step $\Delta t_{\max}$:

$$\Delta t_{\max}^{\text{FIT}} = \frac{1}{c}\left(\sum_u \left(\frac{1}{\Delta u}\right)^2\right)^{-\frac{1}{2}} \quad \text{and} \quad \left(\Delta t_{\max}^{\text{DG}}\right)_{P=0} = 2\Delta t_{\max}^{\text{FIT}} \qquad (3.184)$$

which is called the COURANT-FRIEDRICHS-LEVY (CFL) limit ([77] pp. 214).

For equidistant grids and a homogeneous material distribution in the computational domain $\Omega_{\mathcal{G}}$ this simplifies to:

$$\Delta t_{\max}^{\text{FIT}} = \frac{\Delta u}{c} \quad \text{if } \dim(\Omega_{\mathcal{G}}) = 1 \qquad (3.185)$$

$$\Delta t_{\max}^{\text{FIT}} = \frac{\Delta u}{\sqrt{2}c} \quad \text{if } \dim(\Omega_{\mathcal{G}}) = 2 \qquad (3.186)$$

$$\Delta t_{\max}^{\text{FIT}} = \frac{\Delta u}{\sqrt{3}c} \quad \text{if } \dim(\Omega_{\mathcal{G}}) = 3 \qquad (3.187)$$

The ratio

$$\nu := \frac{c\Delta t}{\Delta u} \qquad (3.188)$$

is called the CFL number ([77] p. 220). A stable time-stepping within the FIT requires $\nu \leq 1$ for the one-dimensional case and $\nu \leq 1/\sqrt{3}$ for the three-dimensional case. For $P = 0$, the DGM requires $\nu \leq 2$ (1D) and $\nu \leq 2/\sqrt{3}$ (3D) for a stable time stepping.

Since these stability limits were derived from the dispersion relation, which itself results from a VON NEUMANN analysis, they are strictly valid only for the conditions of equidistant grids and a homogeneous material distribution. These conditions are commonly not fulfilled for practical applications. In this case, the stability limit is established by a cell-wise evaluation of:

$$\Delta t_{\max}^{\text{FIT}} = \min_i \left\{ \frac{1}{c_i}\left(\sum_{u_i}\left(\frac{1}{\Delta u_i}\right)^2\right)^{-\frac{1}{2}} \right\} \quad \forall i = 1..N \qquad (3.189)$$

with the local speed of light:

$$c_i = \sqrt{\frac{1}{\epsilon_i \mu_i}} \qquad (3.190)$$

For the DGM the maximum time step depends also on the approximation order P. The numerical phase velocity of the BLOCH-modes exceeds the speed of light in continuum (see figs. 3.18 and 3.19). Since all modes are required to be stable, the maximum time step has to be restricted such that

$$\nu^{P\geq 1} = \max\left(c(P)\right)\frac{\Delta t}{\Delta u} \stackrel{!}{\leq} \nu^{P=0} \qquad (3.191)$$

The maximally stable CFL number as a function of P is, hence, given by:

$$\nu_{\max}(P) = \frac{\nu^{P=0}}{\max\left(\Delta u \omega(P)/c\right)} = \frac{2}{\max\left(\Delta u \omega(P)/c\right)} \qquad (3.192)$$

In table 3.1 the maximum CFL number is evaluated for the orders zero to five. It corresponds to the reciprocal value of the maximum of all curves in the respective dispersion graph in the figures 3.18 and 3.19. The values are also plotted in the figure 3.26.

3.6. Consistency, Stability and Convergence

P	0	1	2	3	4	5
ν_{max}	2.0000	0.5000	0.2475	0.1506	0.1016	0.0732

Table 3.1: Maximum stable CFL numbers for the approximation orders P up to five.

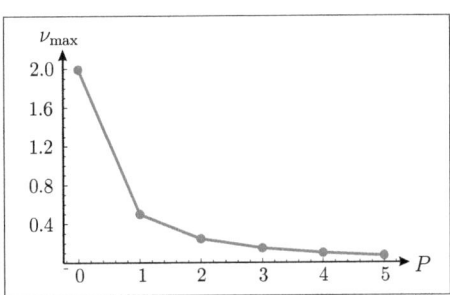

Figure 3.26: Stable CFL numbers for the DGM: For higher approximation orders P the maximally stable CFL number reduces.

The figures 3.27 and 3.28 illustrate the dispersive behavior of the fully discretized FIT and DGM (up to an approximation order of $P = 4$) for waves propagating along an axis of a three-dimensional domain. Figure 3.27 also shows an amplification plot. Both methods are neutrally stable within the stability limit for the applied leap-frog time integration scheme. For higher approximation orders P the phase error of the DGM becomes negligible.

In section 3.5.3 the LT split operator technique was introduced. Its application reduces the three-dimensional problem to a sequence of a two- and a one-dimensional problem. The figures 3.29 to 3.31 show the dispersive behavior of the FIT and the FV methods for one-dimensional problems. For the FV methods also the dissipative behavior is shown. All methods yield exact results for $\nu = 1$. The respective time step is called the *magic time step* Δt^*. This time step is stable for one- but not for three-dimensional problems. The application of the split operator technique, hence, allows for eliminating dispersion and dissipation errors along one coordinate.

The GODUNOV and STRANG splitting are investigated for their stability properties following an alternative approach. In the equations (3.154) and (3.155) on page 63 the time stepping equations for both schemes are given. The respective iteration matrices read:

$$\mathbf{T}^G = \mathbf{T}_T(\Delta t)\, \mathbf{T}_L(\Delta t) \qquad (3.193)$$
$$\mathbf{T}^S = \mathbf{T}_T(\Delta t/2)\, \mathbf{T}_L(\Delta t)\, \mathbf{T}_T(\Delta t/2) \qquad (3.194)$$

It is a necessary and sufficient condition for stability that for the spectral radius ρ of the iteration matrix $\mathbf{T}$:

$$\rho(\mathbf{T}) = \max_i\{|\lambda_i|\} \qquad \forall \lambda_i \in \{\lambda_T\} \qquad (3.195)$$

it holds true [101]:

$$\rho(\mathbf{T}) \leq 1 \qquad (3.196)$$

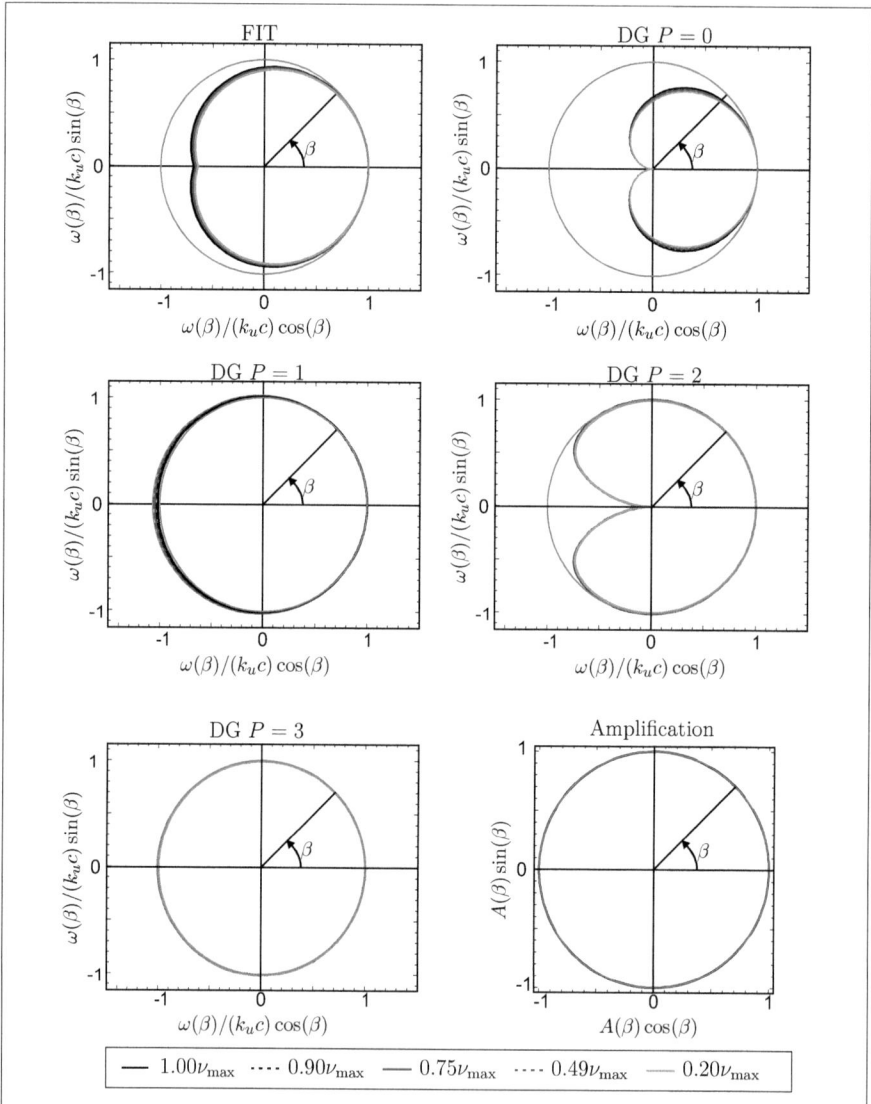

Figure 3.27: Dispersion diagrams of the fully discrete FIT and DGM: Polar plots of the normalized angular frequency of the FIT and the DGM ($P \leq 3$) for waves traveling along a coordinate in a three-dimensional domain. The gray dashed curve is the exact solution. The maximally stable CFL number $\nu_{\max}$ applied in every plot depends on the numerical method and for the DGM also on the approximation order. Additionally, the amplification of the methods is plotted showing their neutral stability within the stability limit.

3.6. Consistency, Stability and Convergence

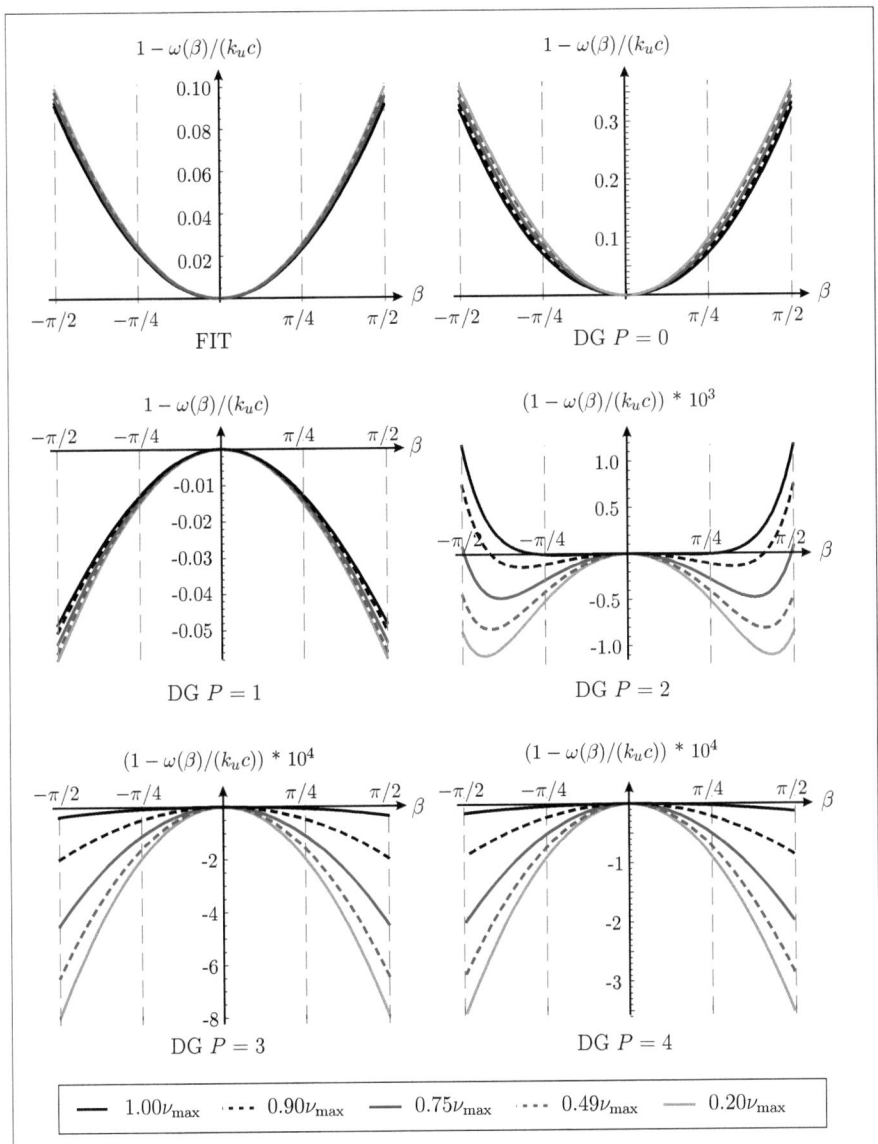

Figure 3.28: Dispersion errors of the fully discrete FIT and DGM: Plots of the error in the numerical angular frequency of the FIT and the DGM for approximation orders of zero to four. The maximally stable CFL number $\nu_{\max}$ applied in every plot depends on the numerical method and for the DGM also on the approximation order.

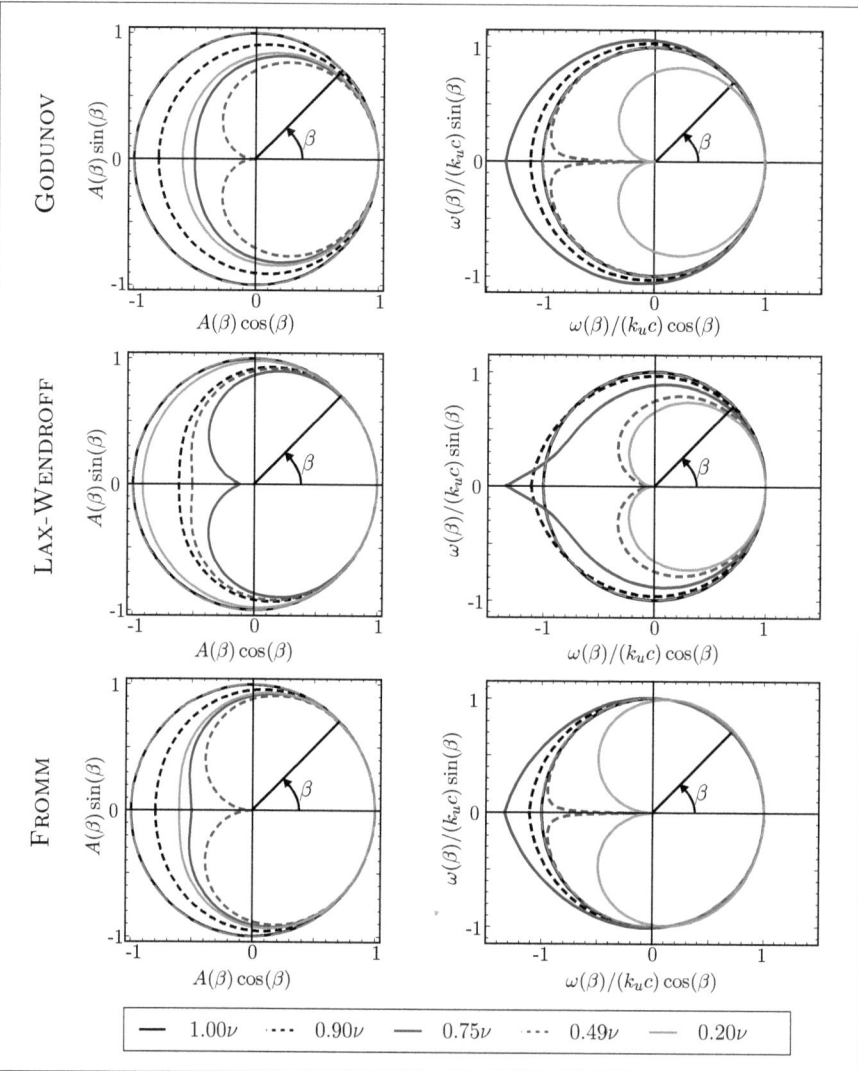

Figure 3.29: Amplification and angular frequency of the GODUNOV, LAX-WENDROFF and FROMM schemes: Polar plots of the amplification (left) and normalized angular frequency (right). The gray dashed curve is the exact solution. GODUNOV's as well as FROMM's method show no phase error for the CFL numbers 1.0 and 0.5. However, for $\nu = 0.5$ damping is at a maximum for both methods.

3.6. Consistency, Stability and Convergence

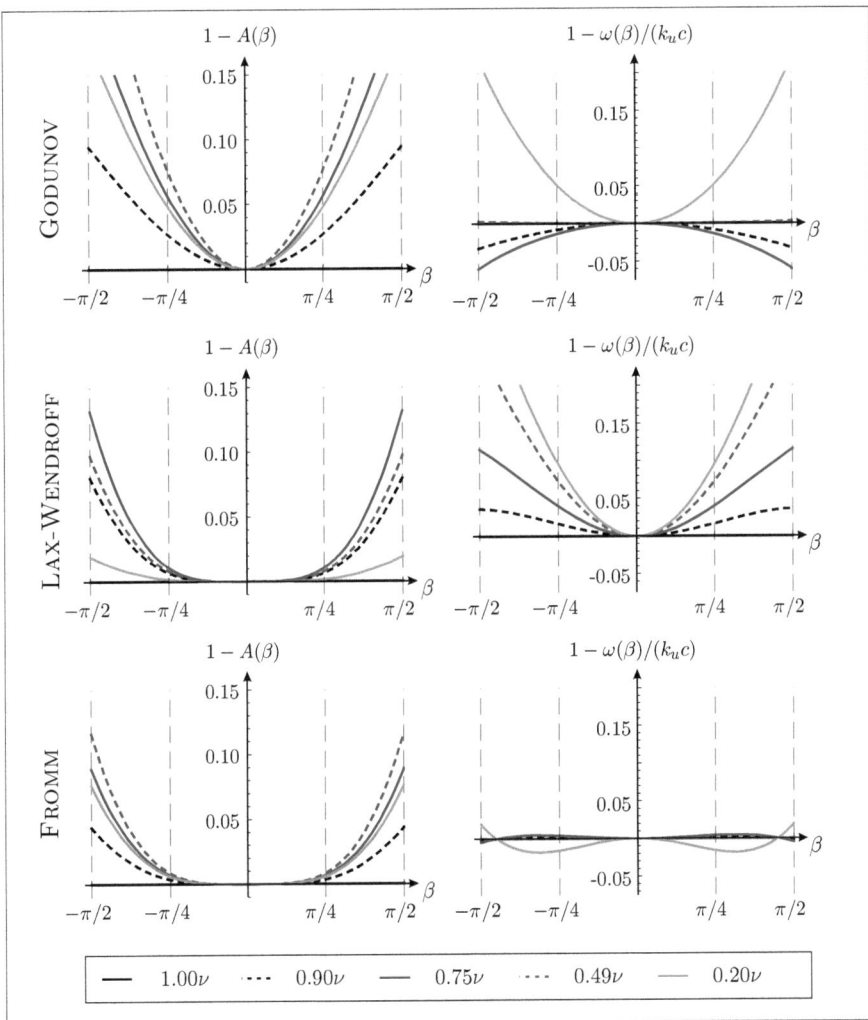

Figure 3.30: Amplitude and angular frequency error of the GODUNOV, LAX-WENDROFF and FROMM schemes: Plots of the amplitude error (left) and normalized angular frequency error (right). The plot ranges are chosen identically for all methods. While GODUNOV's method exposes large amplitude errors, they are small for the LAX-WENDROFF and FROMM methods. GODUNOV's as well as FROMM's method show no phase error for the CFL numbers 1.0 and 0.5. However, for $\nu = 0.5$ damping is at a maximum for both methods. Overall, FROMM's method performs best.

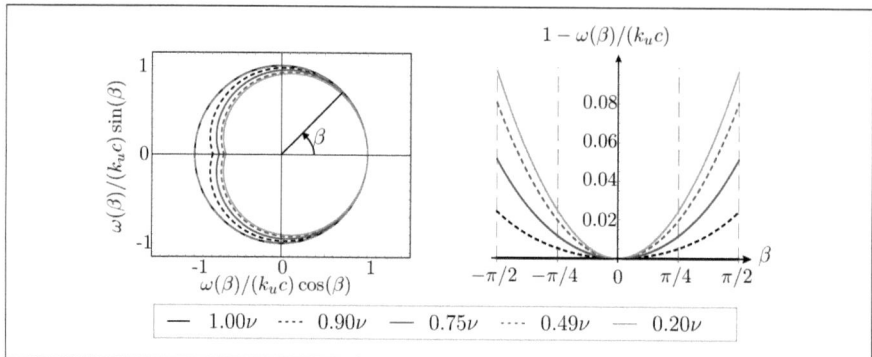

Figure 3.31: Dispersion diagram and dispersion error of the fully discrete FIT for one-dimensional problems: Polar plot of the normalized angular frequency (left) and its error (right). If the maximally stable time step is applied, the dispersion error vanishes.

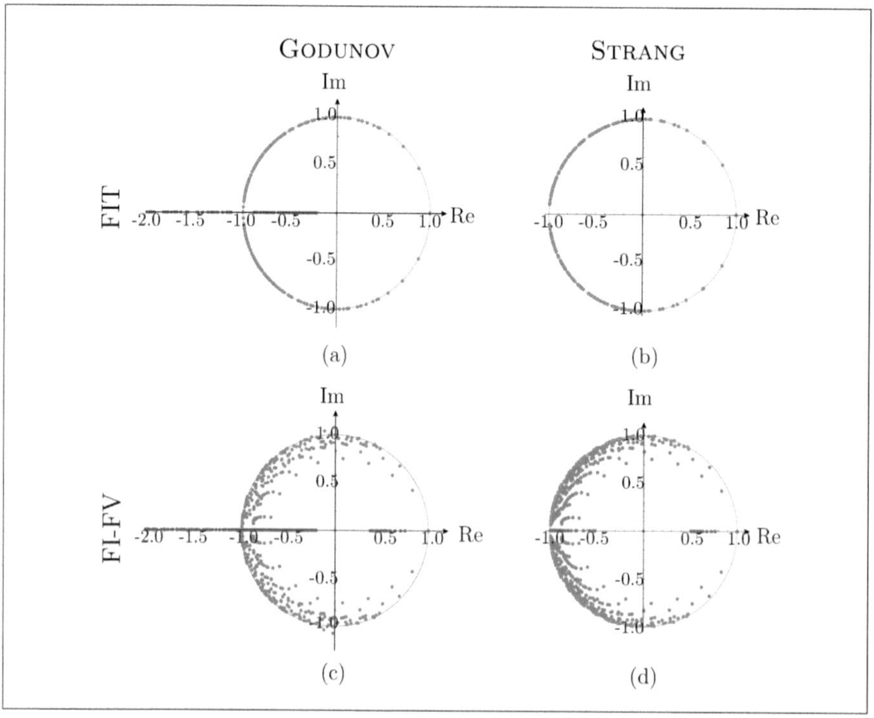

Figure 3.32: Exemplary eigenvalue distribution of the LT-FIT (a,b) and FI-FV (c,d) iteration matrix in the complex plane. In the left column the GODUNOV splitting scheme was applied, while in the right column the STRANG scheme was used. Only the iteration matrices due to the STRANG splitting allow for a stable time stepping.

3.6. Consistency, Stability and Convergence

In figure 3.32 exemplary eigenvalue distributions of $\mathbf{T}^\text{G}$ and $\mathbf{T}^\text{S}$ for the magic time step Δt^* are shown. In (a) and (b) the split operator schemes were applied to the FIT system matrix (3.33), and to the FI-FV system matrix (3.83) in (c) and (d). The former case yields the LT-FIT method, which was established in [80].

For both cases, the GODUNOV splitting violates the stability condition (3.196). The magic time step exceeds the time step limit of the two-dimensional FIT subproblem which is given in (3.186). For the STRANG scheme, the transverse update is decomposed into two steps, each of them proceeding a step of $\Delta t^*/2$ in time. Since it holds true:

$$\frac{\Delta t^*}{2} = \frac{\Delta u}{2c} < \frac{\Delta u}{\sqrt{2}c} = \Delta t_\text{max}^\text{2D} \tag{3.197}$$

the update is stable. The figures 3.32 (b) and (d) confirm this.

Stability limits can also be established from the eigenvalue transformation equation of the respective time integration scheme. For the leap-frog scheme, the RK2 and RK4 methods, and the explicit EULER scheme, these transformations are given in the equations (3.143), (3.149) and (3.162). The time step has to be chosen such that the stability condition (3.196) is fulfilled. However, this requires the knowledge of the largest eigenvalue of the system matrix.

Convergence

The convergence property of a numerical approximation method guarantees that the approximate solution coincides with the true solution in the asymptotic limit. With the previous investigations on consistency and stability at hand, the following theorem allows for stating convergence:

Theorem 3.3 (LAX equivalence theorem). *For a consistent approximation to a well-posed linear evolutionary problem, the stability of the scheme is necessary and sufficient for convergence.*

From this theorem we can immediately follow that all presented methods converge, if the time step is chosen within the stable regime.

It can be noted that the accuracy of a numerical scheme is determined by the specific approximations used. In order to increase the accuracy of the FIT, more accurate material operators are required, for the FVM a more accurate reconstruction of the interface values, and for the DGM a higher order of the approximation space.

3.7 Discrete Charged Particle Dynamics

Up to now the discretization of MAXWELL's equations in space and time was considered. Charged particle dynamics have not been addressed yet. The term *self-consistent charged particle simulations* refers to coupled simulations of MAXWELL's equations and particle dynamics including the effects of the electromagnetic fields excited by the charged particles. These self-fields are called *space charge fields*. Several approaches to the self-consistent simulations are commonly used.

3.7.1 Approaches to Beam Dynamics Simulations

In the following, a short overview of approaches to beam dynamics simulations is given.

The group of the so-called VLASOV *solvers* approach the problem in a direct way by solving the VLASOV equation in time-domain. This equation describes the evolution of the density distribution function of charged particles in the phase space (see sec. 2.2.3) assuming an idealized collisionless plasma [47]. However, self-consistent solutions of the VLASOV equation are computationally very expensive. Currently, only implementations for one- and two-dimensional problems (i.e. in two- and four-dimensional phase space) exist (see e.g. [14, 118]).

A related technique is the *moment approach* to beam dynamics, first proposed in [31]. There, only moments of the phase space distribution are evolved in time (see sec. 2.2.5 for the definition of moments). This greatly reduces the computational burden with respect to VLASOV solvers and, thus, allows for simulations of the six-dimensional phase space. The moment approach is, e.g., implemented in the V-CODE simulation tool [73]. Despite the reduction of the computational costs, the amount of moments to be taken into account increases as $(P+1)^3$ with P the highest regarded order. The amount of moments needed in a simulation becomes, in particular, critical for non-smooth phase space distributions. Nevertheless, programs based on the moment approach make simulations of the beam dynamics in accelerator structures, several meters in length, feasible on a time scale of seconds [120].

Yet another approach to beam dynamics simulations takes the granularity of the phase space distribution into account. Individual particles are traced in phase space. However, even on modern computers it is not (yet) possible to simulate the motion of all particles of a highly charged bunch[5]. Therefore, so-called *macro particles*, carrying the charge and mass of $10^{3..5}$ individual particles, are used in simulations. The equations of motion read:

$$\partial_t \vec{u} = \frac{\vec{u}'}{\gamma} \tag{3.198}$$

$$\partial_t \vec{u}' = \frac{q}{m_0} \left(\vec{E} + \frac{\vec{u}'}{\gamma} \times \vec{B} \right) \tag{3.199}$$

with

$$\vec{u}' = \frac{m(\gamma)\,\vec{v}}{m_0} = \gamma \vec{v} \tag{3.200}$$

(cf. eqns. (2.46) and (2.67)). Thus, the trajectory of a particle depends only on the ratio of its charge and mass at rest q/m_0, allowing for the employment of macro particles. In simulations,

---
[5] An electron bunch carrying a total charge of 1 nC contains approximately 6 bn particles.

3.7. Discrete Charged Particle Dynamics

the particle position and momentum are continuous. The discretization of the temporal derivate ∂_t leads to a second parallel iteration for solving the particle mechanics.

The approaches based on macro particles can be further separated by the applied method for the incorporation of the space charge fields. Some programs, like, e.g., GPT [42] and ASTRA [52] perform a LORENTZ transformation of the particle distribution to the average rest frame of the bunch, where they solve the POISSON equation:

$$\Delta\phi = -\frac{\rho}{\epsilon} \qquad (3.201)$$

The scalar electric potential ϕ is specified by:

$$\vec{E} = -\nabla\phi \qquad (3.202)$$

The field solution is transformed back to the laboratory frame. All programs relying on the solution of this electrostatic equation omit transient field effects. In addition, GPT and ASTRA track the particles in free space, omitting the surrounding structure. However, the space charge fields are scattered at the walls of the structure. These scattered fields are called *wake fields* [135] and their inclusion is a necessity in order to obtain accurate results.

3.7.2 Particle-In-Cell Method

Instead of solving the POISSON equation in the inertial frame, a second approach to the incorporation of space charge effects consists of solving the full MAXWELL's equations in the laboratory frame using convective currents $\vec{J}_c$. These currents are determined from the motion of the macro particles in the computational domain. The evaluation of the discrete field solution at the continuous position of a macro particle for the application of the LORENTZ force in combination with the determination of convective currents due to the particle motion is called the *particle-in-cell* (PIC) algorithm. A survey of PIC methods is given in [16] and more recently in [10].

Convective Current Calculation

FIT: The charge Q of every macro particle is assigned to the surrounding grid nodes according to a *weighting* scheme. The convective currents result from the temporal change of the total charge assigned to the nodes. The simplest particle weighting method is the *nearest grid point* (NGP) scheme. The sum of charges carried by all particles within a dual grid volume $\tilde{\mathcal{G}}_i$ is assigned to the grid node in its center. Using the FIT notation (3.6), this reads:

$$q_i = \sum_j Q_j(\vec{u}) \qquad \forall Q_j \text{ with } \vec{u} \in \tilde{\mathcal{G}}_i \qquad (3.203)$$

Within the NGP scheme the transition of a macro particle from one cell to another is abrupt. Smoother transitions are obtained by distributing the charge Q of a point-like macro particle to a volume in space. This leads to spatially extended particles with a charge density ρ and is referred to as the *cloud-in-cell* (CIC) approach. The amount of charge deposited in a node

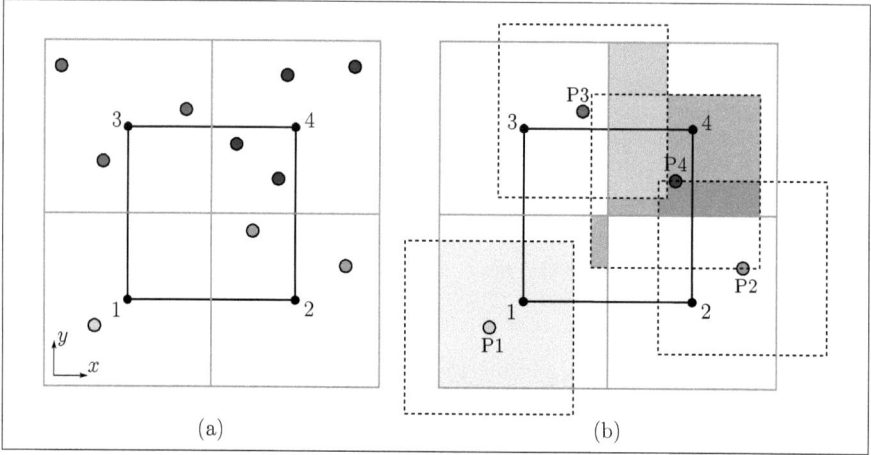

Figure 3.33: Particle weighting schemes: In (a) the *nearest grid point* (NGP) weighting scheme is illustrated. The amount of charge q_i deposited in the grid node i corresponds to the sum of charges in the respective dual volume (dual area in this two-dimensional illustration). In (a) one charge is assigned to node 1, two charges are assigned to node 2 and so on. The *cloud-in-cell* (CIC) weighting is illustrated in (b). The charge of every particle is distributed within a volume equally sized as the grid cells. For every particle the overlapping volume of the charged cloud and the dual cell is calculated according to equation (3.204). For clarity, only the charge deposited to the nodes 1 and 4 is indicated. The charge of the particles P1 and P4 contribute to q_1, and the particles P2, P3 and P4 contribute to the charge q_4 of node 4.

corresponds to the overlapping volume of the dual cell and the particle cloud (see fig. 3.33). In FIT notation, this reads:

$$q_i = \sum_j \int_{\widetilde{\mathcal{G}}_i} \rho_j(\vec{u})\, dV \tag{3.204}$$

Usually, the size of the cloud is chosen to correspond to the size of the grid cells. This results in linear transitions of charges from one cell to another (see fig. 3.34).

For a two-dimensional situation as shown in figure 3.35, the currents are determined by solving:

$$\begin{pmatrix} \Delta q_1 \\ \Delta q_2 \\ \Delta q_3 \\ \Delta q_4 \end{pmatrix} = \begin{pmatrix} -1 & 0 & -1 & 0 \\ 1 & -1 & 0 & 0 \\ 0 & 0 & 1 & -1 \\ 0 & 1 & 0 & 1 \end{pmatrix} \cdot \begin{pmatrix} \widetilde{\jmath}_1 \\ \widetilde{\jmath}_2 \\ \widetilde{\jmath}_3 \\ \widetilde{\jmath}_4 \end{pmatrix} \tag{3.205}$$

where Δq_i is the change of charge deposited in a node in consecutive time steps. The extension of this method to simulations in three-dimensional space is not straightforward since the currents assigned to twelve edges connecting only eight nodes have to be determined. This is addressed in the appendix A.

If the NGP scheme is applied, currents are excited only when particles move from one dual volume to another. This results in peaks of the convective current, which are susceptible to

3.7. Discrete Charged Particle Dynamics

generate numerical noise (see fig. 3.34). In [17] pp. 362 various particle weighting schemes are investigated with regard to their numerical noise properties. In [54] a filtering technique for noise reduction is proposed which was investigated in [79, 108].

FVM and DGM: The determination of convective currents according to the FVM and the DGM follows a different ansatz. The current density is given by:

$$\vec{j}(\vec{r},t) = \rho(\vec{r},t)\,\vec{v}(t) = Q\,\delta(\vec{r}-\vec{r}_\mathrm{p}(t))\,\vec{v}(t) \tag{3.206}$$

where the particle position is $\vec{r}_\mathrm{p}(t)$. The convective current within one time step, thus, reads:

$$\vec{J}_\mathrm{c}(\vec{r},t) = \int_0^{\Delta t} \vec{j}(\vec{r},t)\mathrm{d}t \tag{3.207}$$

This current is projected to the finite element space using the projection operator Π defined in equation (3.97). This yields the FE approximation of the convective current:

$$\vec{\bar{J}}_\mathrm{c} = \Pi \vec{J}_\mathrm{c} = \sum_{i,p} \mathbf{j}_i^p(t)\varphi_i^p(\vec{r}) \tag{3.208}$$

For the FVM only the average current over the volume of a cell has to be evaluated. Here, the macro particles are described as point charges (eqn. (3.206)). It is, however, also possible to choose spatially extended particles, corresponding to the CIC approach [64].

Integration of the Equations of Motion

For the integration of the equations of motion in time the LORENTZ force has to evaluated at the particles positions. Hence, the electric and the magnetic field at the continuous positions have to be determined from the discrete field quantities. In FIT simulations, a trilinear interpolation of the surrounding discrete quantities has to be performed (see fig. 3.36). If the FI-FV method was applied for the spatial discretization, the respective discrete quantity and its slope along the z-coordinate provide a natural way of performing a linear interpolation. Due to the projection of the electromagnetic quantities to a set of basis functions, which are continuous within the extents of a cell, no interpolation has to be performed when applying the DGM. Instead, the approximations given in the equations (3.89) and (3.90) are evaluated at the position $\vec{r}$.

For the simulation of charged particles in accelerator structures, usually the solution of electrostatic (for DC acceleration), eigenmode (for RF acceleration) and magnetostatic problems (for the deflection in magnets) are carried out prior to the beam dynamics simulation. This is necessary in order to determine the exterior electromagnetic fields. In this work, however, an adaptation of the computational grid in time is performed. Due to memory limitations, it is not possible to obtain the exterior fields in the whole computational domain for the maximum grid resolution occurring in the beam dynamics simulation. Thus, a different approach is pursued. Prior to the beam dynamics simulation, all exterior fields are calculated applying the finest feasible grid resolution. Afterwards, the fields along the beam path are extracted and interpolated using a third order spline interpolation. Only the interpolated field data along this path is stored. During the beam dynamics simulation a *paraxial approximation* ([98] pp. 69) at the radial position of each particle is evaluated. A superposition of all exterior fields and the space charge fields yields the electromagnetic field at the particle position. The error committed by applying the paraxial approach is small if the transverse size of the particle bunches is small with respect to the structure.

The integration of the equations of motion cannot be done explicitly since the momentum $\vec{u}'$ appears on the left as well as on the right hand side of equation (3.199). In [66] several so-called LORENTZ *force integrators* are reviewed. The implemented BORIS algorithm decomposes the update of the particle momentum $\vec{u}'$ into three steps. First, the acceleration due to the electric field is performed, proceeding only half a time step $\Delta t/2$. Then, the particle trajectory is rotated according to the present magnetic field. A second half acceleration step completes the momentum update. A detailed description of the algorithm is found in [19] and [66] p. 113. If the BORIS algorithm is applied, the energy of particles in purely magnetic fields is conserved. A verification test of this behavior can be conducted by applying a constant magnetic field, oriented perpendicularly to the direction of motion of a particle. If the time integration method is energy preserving, the particle follows a circular trajectory. In the figure 3.37 the results of this test for the BORIS integrator (a) and the implicit EULER method (b) are illustrated.

A scheme to integrate the set of equations (3.198) and (3.199) is commonly called a *particle pusher*. See also [11, 79, 111] for detailed informations on charged particle simulations using the FIT. In [91] the application of the FVM to particle simulations is reported. The only works, to the knowledge of the author, investigating the coupled problem of MAXWELL's equations and particle dynamics using the DGM are [43, 57, 64, 109].

3.7. Discrete Charged Particle Dynamics

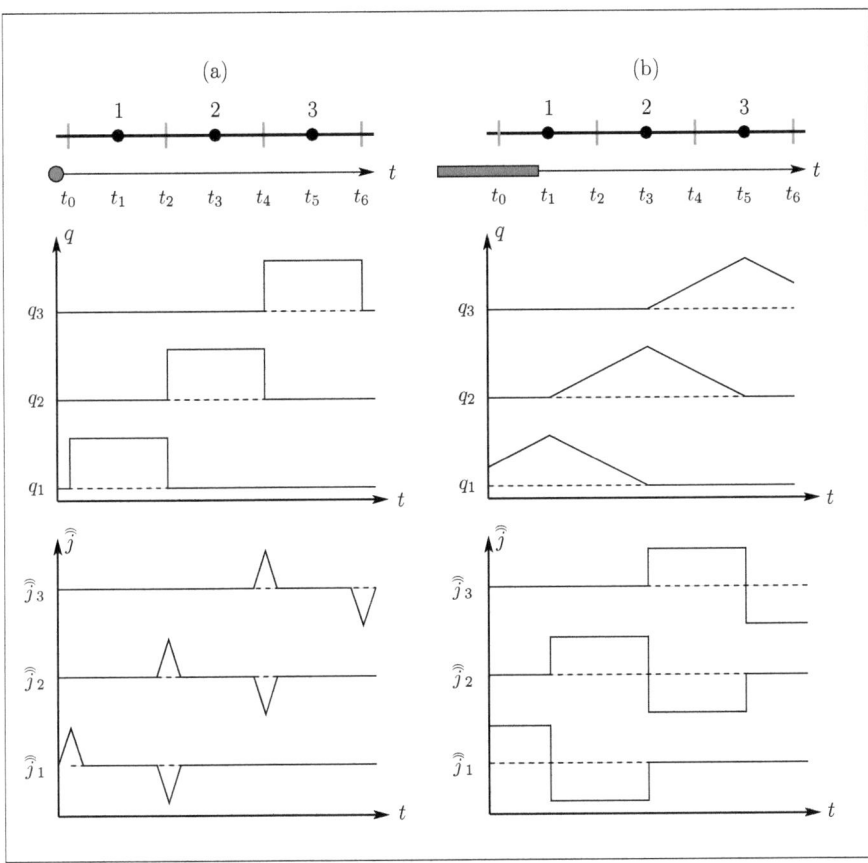

Figure 3.34: FIT charge assignment and grid currents for different particle weighting schemes: The motion of a particle in a one-dimensional grid is illustrated. In the left hand column (a) the charge deposition to the nodes and the resulting grid charges for the NGP scheme is shown. On the right (b), this is illustrated for the CIC scheme, assuming a constant charge density in the charged cloud. The currents equal the temporal derivative of the charges q_n. The temporal profile of the currents due to the NGP scheme (a) corresponds to a series of KRONECKER pulses. The currents resulting from the CIC weighting excite less numerical noise.

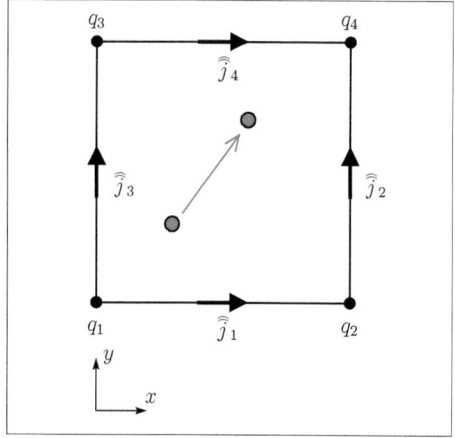

Figure 3.35: PIC method in a two-dimensional cell: The charge of the particle (gray dot) is partially assigned to the surrounding nodes q_i according to the cloud-in-cell method depicted in figure 3.33b. The convective currents $\widehat{\widetilde{j}}_i$ are calculated from the change of the assigned charges with time.

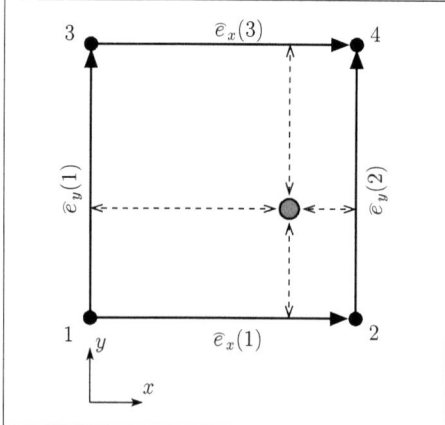

Figure 3.36: Field interpolation to the particle position: When using the FIT, the electric and magnetic field at the position of the particle (gray dot) is determined by a linear interpolation of the surrounding discrete field quantities. In the figure, a bilinear interpolation is indicated. For three-dimensional problems, a trilinear interpolation has to be performed.

3.7. Discrete Charged Particle Dynamics

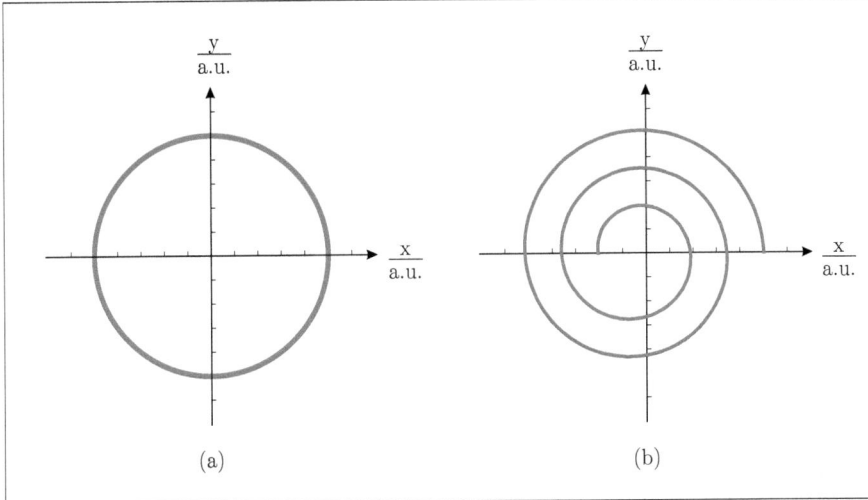

Figure 3.37: Comparison of particle pushers: The BORIS algorithm (a) is an energy conserving integration scheme for the equations of motion (3.198) and (3.199). In the presence of a purely magnetic field oriented perpendicularly to the direction of motion, the particle follows a circular trajectory. In case the integrator is not energy conserving the particle follows a spiraled trajectory. In (b) the implicit EULER method was adopted for the integration.

4. TIME-ADAPTIVE GRID REFINEMENT TECHNIQUES

In the preceding chapter a variety of algorithms for the numerical solution of MAXWELL's equations were introduced. Special attention was paid to the dispersion properties of the methods. The accumulation of dispersion errors severely deteriorates the accuracy of the electromagnetic field solution in long-term simulations. Subsequently, the results of the beam dynamics simulations suffer from numerical noise. In fact, non-physical beam losses induced by numerical noise have been reported [110]. From the dispersion graphs 3.11 (FIT), 3.15 (FVM) and 3.18, 3.19 (DGM) it is evident that reduced dispersion errors can be attained by a reduction of the grid step size Δu and for the DGM also by increasing the approximation order P. Both approaches largely increase the requirements on computer memory and, consequently, the simulation time if they are applied globally in the computational domain. However, a grid refinement in a localized volume can be sufficient for an improved global accuracy of the solution.

In the following section, a short review of existing works and approaches to grid refinement in the FIT, the FVM and the DGM is given. These comprise static as well as time-adaptive grid refinement techniques. Then, a detailed description of the techniques developed for each method is presented in the sections 4.2–4.4. The subject of adaptive grid refinement is closely linked to error estimation. Therefore, the description of the refinement methods is followed by a section dealing with automated time-adaptive hp-refinement for the DGM based on error indicators. In contrast to most publications on this topic, special attention is paid to a fast and computationally inexpensive adaptation strategy.

4.1 Overview of Existing Works

FIT and FDM: Within the framework of the FIT, static non-conformal (see fig. 3.3) local grid refinements, commonly referred to as subgrids, have been developed [88, 112, 129]. This offers the possibility of increasing the solution accuracy, if the limiting factor is given by geometrical details.

The only combination of the FD method with PIC simulations and grid refinement, known to the author, was published by VAY in [132, 133]. It deals with the simulation of heavy ion fusion in static electromagnetic fields. Local grid refinements are performed in dedicated areas like the vicinity of metallic boundaries and the particle emitting surface.

Despite the advantages arising from static subgridding methods in some setups, they offer no benefits if the attainable accuracy of the numerical solution is limited by the accumulation of dispersion errors in long-term simulations. Generally, the simulation of wave propagation phenomena elude the application of grid refinement techniques. The simulation of charged particle bunches in long accelerator structures, however, poses a wave propagation problem, which is highly suitable for the application of a time-adaptive local grid refinement that tracks the particle bunches. Despite the appealing advantages of the approach, it involves largely augmented coding efforts and has been implemented in very few codes. In [44] DEHLER reports on the only implementation of a *zoom function* in the MAFIA code [40] for the accurate simulation of wake fields excited by ultra-short bunches. This approach did not comprise a refined representation of the discretized material distribution (see fig. 3.4). Otherwise, it is equal to the moving window approach, reported in, e.g., [57].

FVM: In the field of Finite Volume Methods, considerable research and development efforts in adaptive grid refinement have been undertaken. FVMs are traditionally applied in fluid- and hydrodynamics rather than in electromagnetics, where shocks and turbulences make the application of local refinements a more pressing matter. Starting from 1989 BERGER, COLELLA and LEVEQUE developed a hierarchical adaptive grid refinement method for two-dimensional applications [12, 13]. It is based on the placement of rectangular *refinement patches*. The size and position of the patches are determined from an error estimator. COHEN et al. made use of wavelet transformations of the discrete solution in order to identify regions containing a large amount of high-frequency spectral components. These regions demand for grid refinement [37]. Adaptive grid refinement for the FVM is still an active area of research.

FEM: In the Finite Element community the pioneering works of BABUSKA on error estimation paved the way for the development of adaptive methods (see [123] and references therein). Among his achievements is the proof of an exponential convergence rate for elliptic problems employing a combination of graded grids (h-refinement) and p-refinement. In [78] LANG reports on the development of an h-adaptive solver for nonlinear parabolic PDE systems with automated, error controlled grid adaptation. It makes use of the so-called *red-green* refinement strategy ([78] p. 60). Since the evaluation of an error estimator is, in general, a computationally expensive operation, an error check is performed only in those cells which experienced a constrained refinement for restoring the regularity of the grid. These cells are labeled as green-refined and during the next grid adaptation step, the error estimator is applied only to them.

An early investigation of p- and hp-refinement was performed by SCHWAB [115]. However, the first adaptive hp-refinement scheme for MAXWELL's equations exhibiting an exponential convergence rate was published by DEMKOWICZ [45]. The first part of the publication dealing with one- and two-dimensional problems appeared in 2007, the second part dealing with general three-dimensional problems in 2008. Based on the ideas of DEMKOWICZ a multi-physics solver was developed, which has also been applied to electrodynamics [117]. This implementation is only about to be extended with capabilities for handling three-dimensional problems [75]. The error estimator is based on the comparison of solutions on multiple grids which makes it computationally expensive. For an extensive overview of development in adaptive FEMs see the introduction in [45].

In the field of particle accelerator simulations CANDEL et al. reported on the implementation of an adaptive p-refinement scheme in combination with the self-consistent simulation of charged particles dynamics [27]. In their implementation, which is restricted to cylindrically symmetric

4.1. Overview of Existing Works

structures, a graded p-refinement is applied within a preset distance from the particle bunch. The reason for applying pure p-refinement is found in the preserved connectivity information of the grid. While this information remains valid in the case of p-refinement, the connectivity information has to be updated if an h-refinement procedure is performed.

DGM: Despite the fact that the method is known since 1973 [97], the Discontinuous GALERKIN Method received considerable attention within the last decade. Following several theoretical investigations of its properties [3, 36, 68, 69], implementations for solving two- and three-dimensional problems in applications arising from various fields including electromagnetics have been reported [28, 94]. Alongside with these publications and benefiting from the knowledge gained from adaptive FV and FE methods, also the development of adaptive DG codes has been reported [74, 99]. In [43, 57] the DGM is employed for performing beam dynamics simulations on statically refined grids. For the numerical solution of MAXWELL's equations an hp-like implementation for two-dimensional problems was published by LANTERI et al. [50]. However, the h-refined areas are determined by geometrical features, like material interfaces, and the approximation order P scales with the size of an element. There is no temporal adaptivity of the refinements.

In this work, only conformal grid refinement is applied, which preserves the regularity of the grid. If a refinement is performed in one cell, all other cells within the cutting are also refined. A review dedicated to structured grid refinement methods is given in [8] and an overview of available software packages performing adaptive grid refinement for various discretization methods is presented in [95]. However, all refinement techniques presented here were implemented from scratch.

4.2 Time-Adaptive Grid Refinement for the Finite Integration Technique

Within the FIT, the adaptation of discrete quantities assigned to the primary and the dual grids has to be distinguished. In figure 4.1 the refinement of a primary grid cell and the subsequent arrangement of electric grid voltages is illustrated. The preservation of the duality of the staggered grids requires augmenting new nodes to the dual grid, as well as shifting existing nodes. This is shown in figure 4.2. For the determination of new or shifted voltages an interpolation has to be carried out. It has to be distinguished between the interpolation of voltages oriented in parallel to the refinement plane (e.g. $\widehat{e}'_x(2')$ in fig. 4.1b) and those oriented perpendicularly (e.g. $\widehat{e}'_z(2)$ and $\widehat{e}'_z(2')$ in fig. 4.1c). For both cases linear and spline interpolations have been investigated.

4.2.1 Interpolation of Grid Voltages

Linear Interpolation

The application of a linear interpolation for the determination of new or shifted voltages is straightforward. For an illustrative presentation, the notation in the following formulae refers to the figures 4.1 and 4.2. As introduced in equation (3.6), c and $\widetilde{c}$ denote the edges of the primary and dual grid. The lengths of the edges are denoted by $|c|$ and $|\widetilde{c}|$. All quantities that are either new or modified during the adaptation procedure are superscripted with a prime.

The values of new electric grid voltages oriented parallel to the refinement plane (e.g. $\widehat{e}'_x(2')$) are determined by:

$$\widehat{e}'_x(2') = \frac{\widehat{e}_x(2) + \widehat{e}_x(3)}{2} \tag{4.1}$$

The splitting of voltages oriented perpendicularly to the refinement plane (e.g. $\widehat{e}_z(2)$) is constrained by the condition:

$$\widehat{e}'_z(2) + \widehat{e}'_z(2') \stackrel{!}{=} \widehat{e}_z(2) \tag{4.2}$$

This preserves the total voltage between the nodes 2 and 3. A detailed sketch of this refinement is shown in figure 4.3. In order to assign values to the refined voltages $\widehat{e}'_z(2)$ and $\widehat{e}'_z(2')$ the voltage gradient along the axis is evaluated using $\widehat{e}_z(1)$ and $\widehat{e}_z(3)$. The refined voltages are given by:

$$\widehat{e}'_z(2) = |c'(2)| \cdot \left(e_z(2) - \frac{|c'(2)|}{2} \cdot \frac{e_z(3) - e_z(1)}{z(\widetilde{3}) - z(\widetilde{1})}\right) \tag{4.3}$$

$$\widehat{e}'_z(2') = |c'(2')| \cdot \left(e_z(2) + \frac{|c'(2')|}{2} \cdot \frac{e_z(3) - e_z(1)}{z(\widetilde{3}) - z(\widetilde{1})}\right) \tag{4.4}$$

where $z(P)$ denotes the z-coordinate of the point P. Furthermore, the sampled electric field

4.2. Time-Adaptive Grid Refinement for the Finite Integration Technique

variables $e_z(i)$ are introduced as:

$$e_z(i) := \frac{\widehat{e}_z(i)}{|c(i)|} \tag{4.5}$$

Since cells are always divided into halves, $|c'(2)|$ equals $|c'(2')|$ and the condition (4.2) is fulfilled.

The interpolation of the magnetic voltages is more cumbersome. Besides the insertion of new dual grid nodes, also existing nodes have to be shifted in position in order to preserve the duality of the staggered grids. Shifting a node, however, imposes a modification of its associated voltages.

The interpolation of magnetic voltages oriented in parallel to the refinement plane is illustrated in figure 4.2b. The voltages $\widehat{h}'_x(\widetilde{6}')$[1] and $\widehat{h}'_x(\widetilde{6}'')$ are given by:

$$\widehat{h}'_x(\widetilde{6}') = \widehat{h}_x(\widetilde{6}) - \left(z(\widetilde{2}) - z(\widetilde{2}')\right) \cdot \frac{\widehat{h}_x(\widetilde{7}) - \widehat{h}_x(\widetilde{5})}{z(\widetilde{3}) - z(\widetilde{2})} \tag{4.6}$$

$$\widehat{h}'_x(\widetilde{6}'') = \widehat{h}_x(\widetilde{6}) + \left(z(\widetilde{2}'') - z(\widetilde{2})\right) \cdot \frac{\widehat{h}_x(\widetilde{7}) - \widehat{h}_x(\widetilde{5})}{z(\widetilde{3})| - z(\widetilde{2})|} \tag{4.7}$$

Similarly to the splitting of electric voltages, the separation of magnetic voltages, oriented perpendicularly to the refinement plane, is constrained by the preservation of the total voltage. However, for a complete description of the refinement algorithm, the bisection of at least two neighboring cells has to be considered. This is illustrated in figure 4.4. For this case, the total voltage in between the nodes $\widetilde{1}$ and $\widetilde{4}$ has to be preserved:

$$\widehat{h}'_z(\widetilde{2}') + \widehat{h}'_z(\widetilde{2}'') + \widehat{h}'_z(\widetilde{3}') + \widehat{h}'_z(\widetilde{3}'') + \widehat{h}'_z(\widetilde{4}) \stackrel{!}{=} \widehat{h}_z(\widetilde{2}) + \widehat{h}_z(\widetilde{3}) + \widehat{h}_z(\widetilde{4}) \tag{4.8}$$

The refined magnetic voltages read:

$$\widehat{h}'_z(\widetilde{2}') = |\widetilde{c}'(\widetilde{2}')| \cdot \left(h_z(\widetilde{2}) - \left(z(2) - \frac{z(\widetilde{1}) + z(\widetilde{2})}{2}\right) \cdot \frac{h_z(\widetilde{4}) - h_z(\widetilde{2})}{z(4) - z(2)}\right) \tag{4.9}$$

$$\widehat{h}'_z(\widetilde{2}'') = \left(z(\widetilde{2}) - z(\widetilde{2}')\right) \cdot h_z(\widetilde{2}) + \left(z(\widetilde{2}'') - z(\widetilde{2})\right) \cdot h_z(\widetilde{3}) \tag{4.10}$$

$$\widehat{h}'_z(\widetilde{3}') = |\widetilde{c}'(\widetilde{3}')| \cdot h_z(\widetilde{3}) \tag{4.11}$$

$$\widehat{h}'_z(\widetilde{3}'') = \left(z(\widetilde{3}) - z(\widetilde{3}')\right) \cdot h_z(\widetilde{3}) + \left(z(\widetilde{3}'') - z(\widetilde{3})\right) \cdot h_z(\widetilde{4}) \tag{4.12}$$

$$\widehat{h}'_z(\widetilde{4}) = |\widetilde{c}'(\widetilde{4})| \cdot \left(h_z(\widetilde{4}) + \left(\frac{z(\widetilde{3}'') + z(\widetilde{4})}{2} - z(4)\right) \cdot \frac{h_z(\widetilde{4}) - h_z(\widetilde{2})}{z(4) - z(2)}\right) \tag{4.13}$$

with the sampled magnetic field $h_z(\widetilde{i})$ defined as:

$$h_z(\widetilde{i}) := \frac{\widehat{h}_z(\widetilde{i})}{|\widetilde{c}(\widetilde{i})|} \tag{4.14}$$

---

[1] Differently from the usual FIT notation, the tilde is also used to mark the index of dual grid points. This makes it easier to distinguish between, e.g., the *new* primary grid point 6' and the *shifted/new* dual grid points $\widetilde{6}'$ and $\widetilde{6}''$.

Since the following holds true:

$$z(2) - \frac{z(\tilde{1}) + z(\tilde{2})}{2} = \frac{z(\tilde{3}'') + z(\tilde{4})}{2} - z(4) \quad (4.15)$$

$$|\tilde{c}'(\tilde{2}')| + \left(z(\tilde{2}) - z(\tilde{2}')\right) = |\tilde{c}(\tilde{2})| \quad (4.16)$$

$$\left(z(\tilde{2}'') - z(\tilde{2})\right) + |\tilde{c}'(\tilde{3}')| + \left(z(\tilde{3}) - z(\tilde{3}')\right) = |\tilde{c}(\tilde{3})| \quad (4.17)$$

$$\left(z(\tilde{3}'') - z(\tilde{3})\right) + |\tilde{c}'(\tilde{4})| = |\tilde{c}(\tilde{4})| \quad (4.18)$$

the equations (4.9)-(4.13) fulfill the condition (4.8). The framed equations define the rules for performing linear interpolations of the state variables $\overline{\mathbf{e}}$ and $\overline{\mathbf{h}}$. They enable grid refinement based on the bisection of cells within the FIT.

Despite the fact that given the rules for grid refinement, grid coarsening within the FIT is straightforward, the respective equations are given for completeness. Consistently with the organization of the grid data in a tree structure, as shown in figure 3.4, the coarsening of the grid follows in the reverse order of its refinement. Thus, only grid nodes inserted in a refinement step can be removed. For multiply refined cells the coarsening order is prescribed by the reverse of the refinement order.

Alongside with the removal of the primary nodes 2' and 6' and the connecting edge, the electric grid voltage $\tilde{e}'_x(2')$ is deleted:

$$\cancel{\tilde{e}'_x(2')} \quad (4.19)$$

No further operation for all electric voltages oriented parallel to the refinement plane is necessary. The constraint (4.2) defines the rule for merging electric voltages oriented perpendicularly to the refinement plane:

$$\tilde{e}_z(2) = \tilde{e}'_z(2) + \tilde{e}'_z(2') \quad (4.20)$$

For the determination of the magnetic coarse grid voltage $\hat{h}_x(\tilde{6})$, oriented parallel to the refinement plane, a linear interpolation of the neighboring voltages allocated on the refined grid is performed:

$$\hat{h}_x(\tilde{6}) = \frac{\hat{h}'_x(\tilde{6}') + \hat{h}'_x(\tilde{6}'')}{2} \quad (4.21)$$

Correspondingly to the electric voltages, the magnetic coarse grid voltages, oriented perpendicularly to the refinement plane, are obtained by the fractional summation of fine grid voltages (see fig. 4.4). The merging of the refined magnetic voltages has to obey the constraint (4.8). The coarse grid voltages are given by:

$$\hat{h}_z(\tilde{2}) = \hat{h}'_z(\tilde{2}') + \left(z(\tilde{2}) - z(\tilde{2}')\right) h'_z(\tilde{2}'') \quad (4.22)$$

$$\hat{h}_z(\tilde{3}) = \left(z(\tilde{2}'') - z(\tilde{2})\right) \cdot h'_z(\tilde{2}'') + \hat{h}'_z(\tilde{3}') + \left(z(\tilde{3}) - z(\tilde{3}')\right) \cdot h'_z(\tilde{3}'') \quad (4.23)$$

$$\hat{h}_z(\tilde{4}) = \left(z(\tilde{3}'') - z(\tilde{3})\right) h'_z(\tilde{3}'') + \hat{h}'_z(\tilde{4}) \quad (4.24)$$

4.2. Time-Adaptive Grid Refinement for the Finite Integration Technique

Since the following holds true:

$$\left(z(\tilde{2}) - z(\tilde{2}')\right) + \left(z(\tilde{2}'') - z(\tilde{2})\right) = |\tilde{c}(\tilde{2}'')| \quad (4.25)$$

$$\left(z(\tilde{3}) - z(\tilde{3}')\right) + \left(z(\tilde{3}'') - z(\tilde{3})\right) = |\tilde{c}(\tilde{3}'')| \quad (4.26)$$

the condition (4.8) is also fulfilled for the case of grid coarsening.

The equations (4.19)-(4.24) describe the manipulations of the FIT field quantities during the coarsening process of the computational grid. In combination with the refinement rules given above, they provide a consistent framework for performing time-adaptive conformal grid refinement within the FIT.

A linear interpolation of the discrete field quantities is easy to implement and fast in code execution. The interpolated quantities are, however, only first order accurate. Especially in regions of strongly varying fields, they do not yield an appropriate representation of high-frequency fields.

Spline Interpolation

Polynomials offer a possibility for performing higher order interpolations. Nevertheless, polynomials of high degrees tend to exhibit an oscillatory behavior and possibly large overshoots which is known as RUNGE's *phenomenon* [104]. Commonly, *spline functions* are employed in order to avoid this.

A spline S is a piecewise defined polynomial function of the order P which is $P-1$ times continuously differentiable, i.e., $S \in \mathcal{C}^{P-1}$[41]. Typically, each piece is of the order of three. The construction rules determine the specific type of spline. If the spline S is demanded to pass exactly through the given data points, and to be twice continuously differentiable (i.e. $S \in \mathcal{C}^2$) with the second derivative equal to zero on every interval boundary, the so-called *natural cubic spline* (C-spline) is obtained. For its determination, a tridiagonal system of equations has to be solved. During a time-domain simulation adopting time-adaptive grid refinement, hundreds to thousands of grid adaptations are performed, each involving thousands of cells. Thus, the solution of a system of equations is computationally too expensive. In addition, the natural cubic spline still exhibits overshoots (see fig. 4.5).

In order to further mitigate RUNGE's phenomenon, the conditions on the continuous differentiability of the spline have to be reduced. A spline S of the order P which is at most $P-2$ times continuously differentiable is called a *broken spline* or *subspline* ([130] p. 464). The AKIMA spline is a cubic subspline which is an element of the functional space $\mathcal{C}^1$ [5]. Cubic (sub-)splines can conveniently be characterized by a triple of values for each data point i:

$$(x_i, f_i, s_i) \quad (4.27)$$

with x_i the coordinate of the data point, f_i its value and s_i its slope. Imposing the conditions:

$$S(x_{i-1}) = f_{i-1} \qquad S(x_i) = f_i \quad (4.28)$$
$$S'(x_{i-1}) = s_{i-1} \qquad S'(x_i) = s_i \quad (4.29)$$

at every interval boundary, uniquely describes the broken cubic spline function:

$$S(x) = a_0 + a_1(x - x_{i-1}) + a_2(x - x_{i-1})^2 + a_3(x - x_{i-1})^3 \quad (4.30)$$

with the coefficients:

$$a_0 = f_{i-1} \tag{4.31}$$
$$a_1 = s_{i-1} \tag{4.32}$$
$$a_2 = \frac{3}{(\Delta x)^2}(f_i - f_{i-1}) - \frac{1}{\Delta x}(s_i + 2s_{i-1}) \tag{4.33}$$
$$a_3 = \frac{2}{(\Delta x)^3}(f_i - f_{i-1}) + \frac{1}{(\Delta x)^2}(s_i + s_{i-1}) \tag{4.34}$$

However, the slopes s_i are, in general, unknown. The AKIMA procedure makes use of a local heuristic method for the estimation of the slopes. It involves the data point i and two neighboring points on each side. First, the piecewise gradients g, given by:

$$g_j := \frac{f_{j+1} - f_j}{x_{j+1} - x_j} \quad \text{with} \quad j = i-2, i-1, i, i+1 \tag{4.35}$$

are evaluated. These are weighted with the factors w:

$$w_{i-1} := |g_{i+1} - g_i| \tag{4.36}$$
$$w_i := |g_{i-1} - g_{i-2}| \tag{4.37}$$

yielding the estimated slope at the data point i:

$$s_i := \frac{w_{i-1} g_{i-1} + w_i g_i}{w_{i-1} + w_i} \tag{4.38}$$

The fundamental idea of the heuristic is to assign a larger weighting factor to, e.g., the right hand side gradient (g_i) of the data point i if the difference between the two gradients on the left hand side (g_{i-1} and g_{i-2}) is small and vice versa. This is intended to minimize overshooting and oscillatory behavior.

AKIMA's heuristic is closely related to the idea of *slope limiting*. Slope limiters were introduced by VAN LEER [131] for the reconstruction of interface values in Finite Volume Methods (see sec. 3.3.2 pp. 42). Using the *minmod limiter* ([83] p. 186) the estimated slope reads:

$$s_i := \text{minmod}(g_i, g_{i+1}) = \begin{cases} g_i & \text{if } |g_i| < |g_{i+1}| \text{ and } g_i g_{i+1} > 0 \\ g_{i+1} & \text{if } |g_i| > |g_{i+1}| \text{ and } g_i g_{i+1} > 0 \\ 0 & \text{if } g_i g_{i+1} \leq 0 \end{cases} \tag{4.39}$$

Out of its two arguments, the minmod operator chooses the smaller one if they are equally signed and zero otherwise. An overview of slope limiting techniques is found in [77, 83].

In figure 4.5 the interpolation of a set of data points using linear interpolation, the C-spline and the AKIMA spline are illustrated. For this rather artificial case, the C-spline shows an oscillatory behavior and strong overshooting. For the determination of the AKIMA spline up to data point 10, the slope was calculated using AKIMA's slope definition (4.38). From this point on, the minmod limiter was applied, which effectively avoids any overshooting.

The actual spline interpolation procedure of the discrete field quantities is illustrated in figure 4.6. In (a) and (b) the dual grid points along the z-coordinate are shown. In (b) the refined area has moved by a distance of one grid cell into the z-direction. Therefore, the grid topology

4.2. Time-Adaptive Grid Refinement for the Finite Integration Technique

in the cells 2 and 5 has to be modified. The spline is set up using the position of the grid points and the allocated field data in the refined area and some neighboring cells. An evaluation of the spline function at the position of new or shifted nodes yields the interpolated field values.

The AKIMA and the slope limited splines are viable choices for performing interpolations within the computational grid. They are third order accurate and constructed from local information. Thus, they do not require the solution of a system of equations. Consequently, the setup is computationally fast and occupies a negligible amount of computer memory.

4.2.2 Stability

The incorporation of time-adaptive grid refinement with the FIT cannot a priori be expected to preserve the stability of the algorithm. Since, however, the regularity of the grid is maintained in every adaptation, the duality relation holds true for every time step n:

$$\mathbf{C}^n = (\tilde{\mathbf{C}}^n)^\mathrm{T} \qquad \forall n \in [1, N] \tag{4.40}$$

where $\mathbf{C}^n$ and $\tilde{\mathbf{C}}^n$ are the time-dependent curl matrices. As shown in the subsections 3.2.2 (pp. 33) and 3.5.1 (pp. 58), this is a necessary and sufficient condition to guarantee energy conservation and, thus, the stability of the time stepping. Consequently, it is sufficient to find a norm of the discrete electromagnetic field energy which is non-increasing during the interpolation procedure in order to demonstrate the stability of the adaptive method.

The usual definition (3.27) does not serve this purpose. It is easily demonstrated that due to the constraints (4.2) and (4.8) the norm is not preserved during grid adaptations. Calculating the contribution of the voltage $\widehat{e}_z(2)$ of figure 4.1 before and after refinement yields:

$$M_{\epsilon,z}(2) \cdot (\widehat{e}_z(2))^2 = M_{\epsilon,z}(2) \cdot [\widehat{e}'_z(2) + \widehat{e}'_z(2')]^2 \neq M_{\epsilon,z}(2) \cdot \left[(\widehat{e}'_z(2))^2 + (\widehat{e}'_z(2'))^2\right] \tag{4.41}$$

Since a continuous refinement of the computational grid ultimately leads to a continuous field description, the idea is to define a space-continuous energy density based on the space-discrete FIT quantities. In [114], an energy for the extent of each cell is defined by means of an averaging process of primal and dual quantities. This leads to piecewise constant energy densities throughout the grid. This definition is consistent and coincides with the continuous energy density defined in (2.20) for $\Delta u \rightarrow 0$ ([49] p. 41). Nevertheless, it is directly linked to the number of grid cells.

Instead, a linear interpolation operator $\mathcal{I}$ is introduced in order to define a space-continuous energy density $\overline{w}(\vec{r})$ based on the sampled discrete field quantities:

$$\overline{w}(\vec{r}) := \frac{1}{2}\left(\epsilon(\vec{r}) \cdot (\mathcal{I}(\mathbf{e},\vec{r}))^2 + \mu(\vec{r}) \cdot (\mathcal{I}(\mathbf{h},\vec{r}))^2\right) \tag{4.42}$$

The interpolation operator is defined as:

$$\mathcal{I}(\mathbf{e},\vec{r}) := \{\mathcal{I}_x(\mathbf{e}_x,\vec{r}),\mathcal{I}_y(\mathbf{e}_y,\vec{r}),\mathcal{I}_z(\mathbf{e}_z,\vec{r})\} \tag{4.43}$$

with

$$\begin{aligned}\mathcal{I}_x(\mathbf{e}_x,\vec{r}) := {}& e_x(i_x,i_y,i_z) \\ & + \frac{e_x(i_x,i_y+1,i_z) - e_x(i_x,i_y,i_z)}{y(i_y+1) - y(i_y)} \cdot (y - y(i_y)) \\ & + \frac{e_x(i_x,i_y,i_z+1) - e_x(i_x,i_y,i_z)}{z(i_z+1) - z(i_z)} \cdot (z - z(i_z))\end{aligned} \tag{4.44}$$

and the indices (i_x, i_y, i_z) such that $\vec{r} \in V(i_x, i_y, i_z)$. The operators $\mathcal{I}_y(\mathbf{e}_y, \vec{r}), \mathcal{I}_z(\mathbf{e}_z, \vec{r})$ and $\mathcal{I}(\mathbf{h}, \vec{r})$ are defined accordingly. This linear interpolation operator allows for a continuous description of the electric and magnetic field throughout the computational domain. It is obviously the general interpolation framework which includes the afore presented calculation of grid voltages for bisection based grid adaptation.

Next, the energy norm $\overline{\mathcal{W}}$ is defined as the volume integral over the space-continuous energy density $\overline{w}(\vec{r})^2$:

$$\overline{\mathcal{W}} = \int_{\Omega_\mathcal{G}} \overline{w}(\vec{r}) \, dV \tag{4.45}$$

It is sufficient to consider the refinement of one cell, as illustrated in figure 4.7. If the energy $\overline{\mathcal{W}}$ is non-increasing during one refinement step of this cell, it is also non-increasing for a repetitive bisection procedure. Using the notation of figure 4.7, the energy is given by:

$$\begin{aligned}
\overline{\mathcal{W}}^{\mathrm{I}} &= \int_{\Delta x} \int_{\Delta y} \int_{z(1)}^{z(2)} dx\,dy\,dz\, \overline{w}(\vec{r}) \\
&= \frac{1}{2} \int_{\Delta x} \int_{\Delta y} \int_{z(1)}^{z(2)} dx\,dy\,dz \left(\epsilon(\vec{r}) \cdot (\mathcal{I}(\mathbf{e}^{\mathrm{I}}, \vec{r}))^2 + \mu(\vec{r}) \cdot (\mathcal{I}(\mathbf{h}^{\mathrm{I}}, \vec{r}))^2 \right) \\
&= \frac{1}{2} \left(\int_{\Delta x} \int_{\Delta y} \int_{z(1)}^{z(1')} dx\,dy\,dz \left(\epsilon(\vec{r}) \cdot (\mathcal{I}(\mathbf{e}^{\mathrm{I}}, \vec{r}))^2 + \mu(\vec{r}) \cdot (\mathcal{I}(\mathbf{h}^{\mathrm{I}}, \vec{r}))^2 \right) \right. \\
&\quad \left. + \int_{\Delta x} \int_{\Delta y} \int_{z(1')}^{z(2)} dx\,dy\,dz \left(\epsilon(\vec{r}) \cdot (\mathcal{I}(\mathbf{e}^{\mathrm{I}}, \vec{r}))^2 + \mu(\vec{r}) \cdot (\mathcal{I}(\mathbf{h}^{\mathrm{I}}, \vec{r}))^2 \right) \right) \\
&= \frac{1}{2} \int_{\Delta x} \int_{\Delta y} \int_{z(1)}^{z(1')} dx\,dy\,dz \left(\epsilon(\vec{r}) \cdot (\mathcal{I}(\mathbf{e}^{\mathrm{II}}, \vec{r}))^2 + \mu(\vec{r}) \cdot (\mathcal{I}(\mathbf{h}^{\mathrm{II}}, \vec{r}))^2 \right) \\
&\quad + \frac{1}{2} \int_{\Delta x} \int_{\Delta y} \int_{z(1')}^{z(2)} dx\,dy\,dz \left(\epsilon(\vec{r}) \cdot (\mathcal{I}(\mathbf{e}^{\mathrm{II}}, \vec{r}))^2 + \mu(\vec{r}) \cdot (\mathcal{I}(\mathbf{h}^{\mathrm{II}}, \vec{r}))^2 \right) \\
&= \overline{\mathcal{W}}^{\mathrm{II}}
\end{aligned} \tag{4.46}$$

The above equality holds true since:

$$e^{\mathrm{II}}(1') = \mathcal{I}\left(\mathbf{e}^{\mathrm{I}}, \vec{r}(1')\right) \quad \text{and} \quad e^{\mathrm{II}}(3') = \mathcal{I}\left(\mathbf{e}^{\mathrm{I}}, \vec{r}(3')\right) \tag{4.47}$$

or more generally:

$$\mathcal{I}(\mathbf{e}^{\mathrm{I}}, \vec{r}) \equiv \mathcal{I}(\mathbf{e}^{\mathrm{II}}, \vec{r}) \quad \text{and} \quad \mathcal{I}(\mathbf{h}^{\mathrm{I}}, \vec{r}) \equiv \mathcal{I}(\mathbf{h}^{\mathrm{II}}, \vec{r}) \quad \forall \vec{r} \in \Omega_\mathcal{G} \tag{4.48}$$

It follows that the energy of the discrete electromagnetic field, specified by the norm $\overline{\mathcal{W}}$ (4.45), with the space-continuous energy density $\overline{w}$ (4.42), is a conserved quantity for the grid adaptation algorithm based on linear interpolation. The identities (4.48) represent the requirements,

---

[2] In order to show that $\overline{\mathcal{W}}$ is a norm, eqn. (4.42) is rewritten as $\overline{w}(\vec{r}) = \frac{1}{2}(\epsilon(\vec{r}) \cdot \bar{e}(\vec{r})^2 + \mu(\vec{r}) \cdot \bar{h}(\vec{r})^2)$. Since ϵ and μ are strictly positive, $\overline{\mathcal{W}}$ is always a positive number unless the space-continuous electric and magnetic fields $\bar{e}(\vec{r})$ and $\bar{h}(\vec{r})$ are zero for every $\vec{r} \in \Omega_\mathcal{G}$ in which case $\overline{\mathcal{W}}$ is zero.

which have to be fulfilled in order for the norm $\overline{\mathcal{W}}$ to be conserved during grid adaptation. The interpolating function has to be unique and is not allowed to alter if further consistent data points are added within the computational domain. However, these identities are also fulfilled if the linear interpolation operator (4.44) is replaced with the spline interpolation (4.30). This immediately allows for stating the stability of the grid adaptation based on the presented spline interpolation.

The stability of the grid adaptation algorithms has been validated in numerous numerical experiments.

The application of a linear interpolation operator for the definition of a space-continuous description based on the discrete field quantities is closely related to an idea presented in [113]. There, the FIT quantities are reinterpreted by means of WHITNEY basis functions. This allows for an evaluation of the discrete quantities at any position within the computational domain (cf. eqns. (7) and (11) in [113]). The basic idea of applying the WHITNEY-FEM on staggered grids in order to synthesize the FIT and FEM approaches was proposed in [22] before. In [49] pp. 183 the equivalence of the lowest order WHITNEY-FEM formulation and the FIT is proven.

4.2.3 Charged Particle Dynamics on Adaptive Grids

In section 3.7.2, the calculation of convective currents in the computational domain was described. For the cloud-in-cell approach the charge Q is distributed within a volume in space. The extent of this volume is usually chosen to coincide with the size of the cells. However, if a non-equidistant grid is employed or in the case of adaptive grid refinement, the non-uniform sizes of the cells demand for a modification of the algorithm.

There are two modification options: First, a constant size of the particle cloud is chosen independently from the grid cell size or, second, the size of the cloud is adapted in order to match the sizes of the involved cells. Depending on the local level of grid refinement, the first option can largely increase the number of cells affected by one charged particle cloud. Besides the coding efforts coming along with this non-local operation, the deposit of fractional charges to a larger number of grid points (see fig. 3.33b) increases the computational load. Therefore, the second option has been implemented. The adaptation of the cloud in dependence of the grid cell size is illustrated in figure 4.8 for a one-dimensional example.

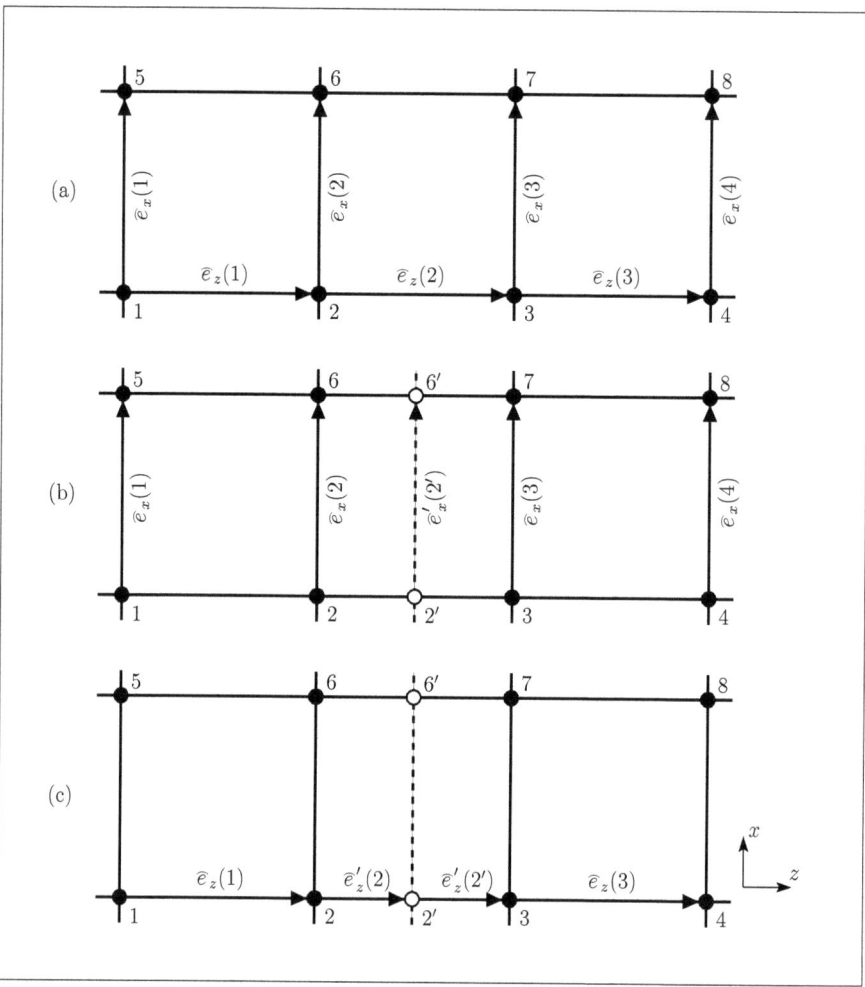

Figure 4.1: Grid adaptation within the FIT (Allocation of electric grid voltages): This figure illustrates the refinement of the primary grid and the associated interpolations of electric grid voltages in the $x-z$ plane. In (a) the initial situation is shown. The arrows indicate the electric voltages allocated to the respective edges. In (b) and (c) the middle grid cell was refined, introducing an additional grid line (dashed). Thus, the two new nodes 2' and 6' (circles) are inserted. All voltages that are new or require an update of their value are marked with a prime ($\widehat{e}'$). In (b) the position of the new electric grid voltage $\widehat{e}'_x(2')$ is shown. The voltage $\widehat{e}_z(2)$, oriented along the z-coordinate, has to be split into the two voltages $\widehat{e}'_z(2)$ and $\widehat{e}'_z(2')$ shown in (c).

4.2. Time-Adaptive Grid Refinement for the Finite Integration Technique

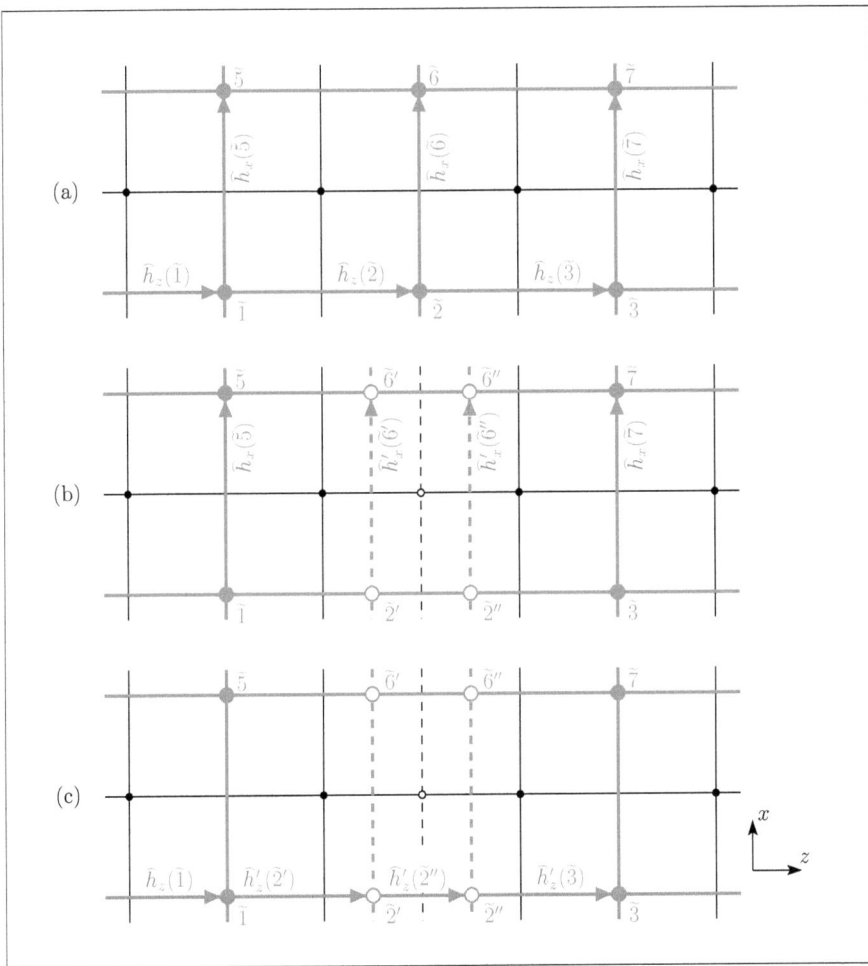

Figure 4.2: Grid adaptation within the FIT (Allocation of magnetic grid voltages): This figure illustrates the refinement of the dual grid (gray), imposed by the refined primary grid (thin black lines). The initial situation is shown in (a). In order to preserve the duality of the two grids (see fig. 3.9) the dual nodes $\widetilde{2}$ and $\widetilde{6}$ have to be shifted. Their new position is indicated by $\widetilde{2}'$ and $\widetilde{6}'$ (b and c). In addition, the two new nodes $\widetilde{2}''$ and $\widetilde{6}''$ have to be inserted. The dashed gray lines show the position of the shifted and the new dual edges. On the dual grid, two interpolations have to be performed in order to determine the values of the x-directed magnetic voltages $\widehat{h}'_x(\widetilde{6}')$ and $\widehat{h}'_x(\widetilde{6}'')$ shown in (b). The z-oriented voltages $\widehat{h}_z(\widetilde{2})$ and $\widehat{h}_z(\widetilde{3})$ have to be split into the voltages $\widehat{h}'_z(\widetilde{2}')$, $\widehat{h}'_z(\widetilde{2}'')$ and $\widehat{h}'_z(\widetilde{3})$ (c).

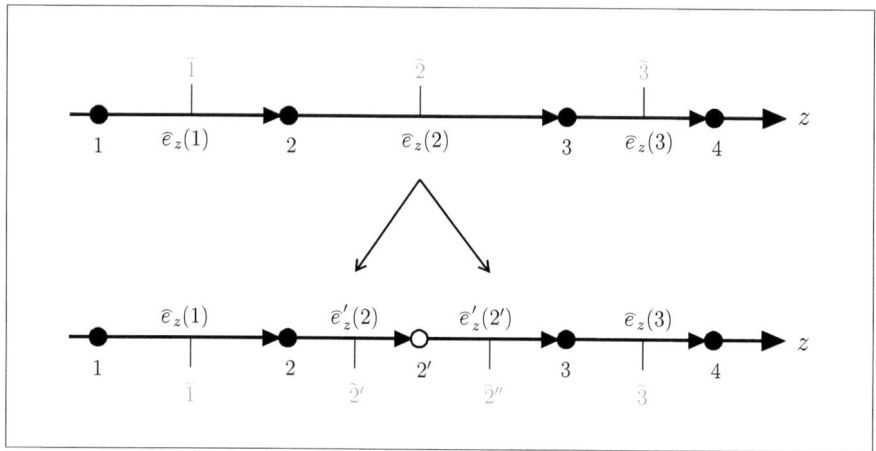

Figure 4.3: Interpolation of electric grid voltages oriented perpendicularly to the refinement plane: The coarse grid voltage $\widehat{e}_z(2)$ is split into the two voltages, $\widehat{e}'_z(2)$ and $\widehat{e}'_z(2')$, such that the total voltage in between the points 2 and 3 is preserved (eqn. (4.2)). In order to assign values to $\widehat{e}'_z(2)$ and $\widehat{e}'_z(2')$ the voltage gradient along the axis is evaluated using $\widehat{e}_z(1)$ and $\widehat{e}_z(3)$ (eqns. (4.3), (4.4)).

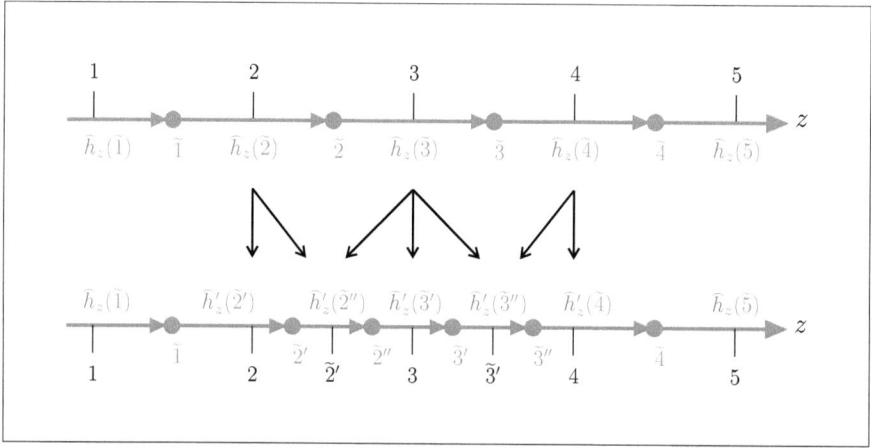

Figure 4.4: Interpolation of magnetic grid voltages oriented perpendicularly to the refinement plane: The voltages $\widecheck{h}_z(\widetilde{2})$, $\widecheck{h}_z(\widetilde{3})$ and $\widecheck{h}_z(\widetilde{4})$, assigned to coarse grid edges, are split and distributed such that the total magnetic voltage in between the points $\widetilde{2}$ and $\widetilde{4}$ is preserved (eqn. (4.8)). The voltages on the refined grid are determined according to the equations (4.9)-(4.13).

4.2. Time-Adaptive Grid Refinement for the Finite Integration Technique

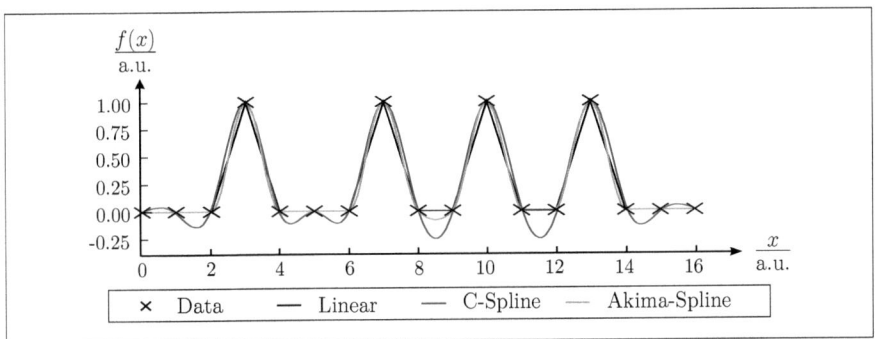

Figure 4.5: Comparison of spline interpolations: The given data points are interpolated using linear interpolation, a natural cubic spline (C-spline) and an AKIMA spline. For this (artificial) data, the C-spline exhibits an oscillatory behavior and strong overshooting. These effects can be mitigated, if the demands on the continuous differentiability of the spline are reduced. The AKIMA spline is a cubic subspline. Up to the data point 10, AKIMA's original definition was employed for the estimation of the slope (eqn. (4.38)) at the position of the data points. From this point on the minmod slope limiter (4.39) was applied, which effectively avoids overshooting in between the data points 11 and 12.

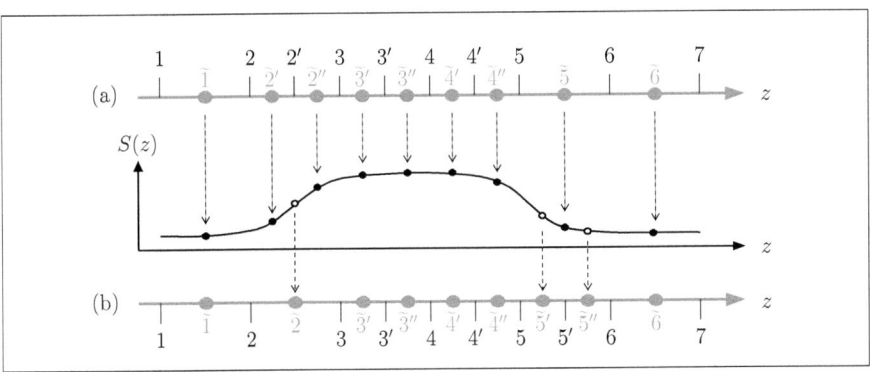

Figure 4.6: Setup and evaluation of the interpolating spline: An arrangement of dual grid points along the z-coordinate is shown in (a) and (b). In (b) the refined area of the grid has moved by a distance of one grid cell into the z-direction, requiring the modification of the grid topology. The interpolating spline is set up, using the grid points and field data in the refined area and some neighboring cells (black dots). An evaluation of the spline function at the position of new or shifted nodes (circles), yields the interpolated field values.

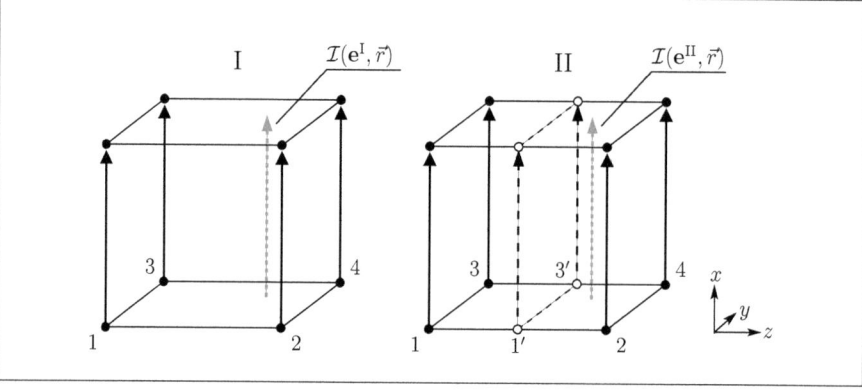

Figure 4.7: Determination of continuous sampled quantities: A space-continuous description of the electromagnetic fields can be obtained by a linear interpolation of the surrounding space-discrete grid voltages. The interpolation operator $\mathcal{I}$ (eqns. (4.43), (4.44)) forms the general interpolation framework which allows for the determination of the respective field component at any position within the cell (indicated in gray). This includes the determination of voltages on refined edges, e.g., $\widehat{e}_x(1')$ and $\widehat{e}_x(3')$ in II, as a special case. If the conditions (4.48) are fulfilled, the values of $\mathcal{I}(\mathbf{e}^\mathrm{I}, \vec{r})$ and $\mathcal{I}(\mathbf{e}^\mathrm{II}, \vec{r})$ are identical.

4.2. Time-Adaptive Grid Refinement for the Finite Integration Technique

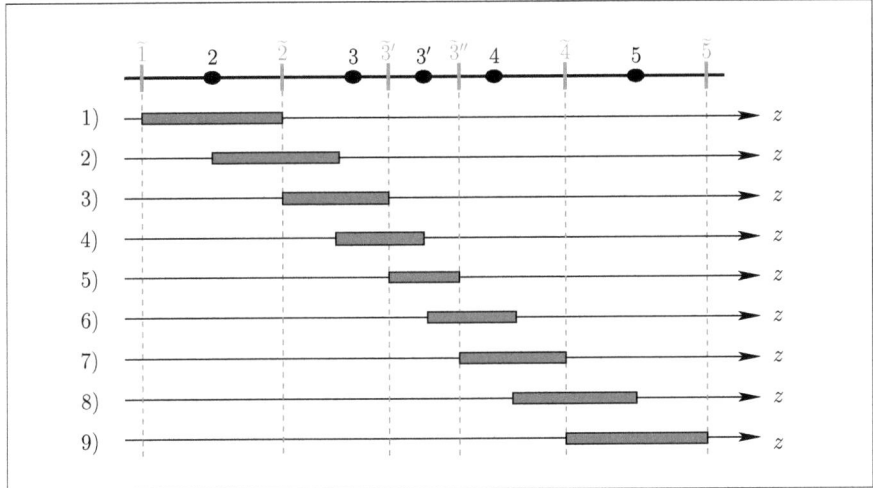

Figure 4.8: Illustration of the adaptation of the particle cloud size in one dimension: The size of the particle cloud relates to the extent of the covered cells on the dual grid. If the particle is centered within a dual grid cell (cases 1, 3, 5, 7 and 9) the sizes of the cloud and the dual volume coincide. Otherwise, the cloud has to be adapted asymmetrically around the particle position. For the cases 2 and 4 the charge density within the part of the cloud within the dual volumes $\tilde{2}$ or $\tilde{3}'$ has to be increased as a consequence of the diminished size. Contrarily, for the cases 6 and 8, the charge density of the cloud in the dual volumes $\tilde{3}''$ and $\tilde{4}$ has to be decreased.

4.3 Time-Adaptive Grid Refinement for the Finite Integration-Finite Volume Method

In the section 3.3 the hybrid FI-FV method was introduced and investigated. The Finite Volume Method does not employ a staggered grid doublet but only one grid for the discretization of all components of the electromagnetic field. Within the hybrid method, the FIT is applied in the $x-y$-plane while the FVM operates along the z-direction. However, the stored DoF are those obtained by the FIT discretization procedure. They are only temporarily converted to FV state variables by applying the transformation operator $\mathbf{T}$ given in (3.88). Hence, it is sufficient to consider the interpolation of FIT variables. The lack of a dual grid in the direction of grid refinement largely simplifies the interpolation of the discrete quantities. The sketches shown in the figures 4.1 and 4.3 are sufficient to describe all interpolations. In the equations (4.1), (4.3) and (4.4) the linear interpolation of the electric voltages is described. For the non-staggered grid setup along the z-direction, the respective equations also apply for interpolating the magnetic grid voltages:

$$\widehat{h}'_x(2') = \frac{\widehat{h}_x(2) + \widehat{h}_x(3)}{2} \tag{4.49}$$

$$\widehat{h}'_z(2) = |c'(2)| \cdot \left(h_z(2) - \frac{|c'(2)|}{2} \cdot \frac{h_z(3) - h_z(1)}{z(\widetilde{3}) - z(\widetilde{1})} \right) \tag{4.50}$$

$$\widehat{h}'_z(2') = |c'(2')| \cdot \left(h_z(2) + \frac{|c'(2')|}{2} \cdot \frac{h_z(3) - h_z(1)}{z(\widetilde{3}) - z(\widetilde{1})} \right) \tag{4.51}$$

The notation $z(\widetilde{P})$ is used for brevity and indicates the intermediate point of $z(P)$ and $z(P+1)$, as shown in figure 4.3. In accordance with (4.5), the sampled magnetic field $h_z(i)$ is given by:

$$h_z(i) := \frac{\widehat{h}_z(i)}{|c(i)|} \tag{4.52}$$

For obtaining the electric grid voltages during grid coarsening, the equations (4.19) and (4.20) are applied. The respective equations for the magnetic voltages read:

$$\widehat{h}'_x(2') \tag{4.53}$$

$$\widehat{h}_z(2) = \widehat{h}'_z(2) + \widehat{h}'_z(2') \tag{4.54}$$

The application of a spline interpolation follows the ideas presented in the preceding section. The setup of the spline function simplifies since the z-positions of the supporting, as well as the evaluation points are identical for all electric and magnetic field components.

Since the discrete electromagnetic variables within the FIT and the FI-FV method are identical, the proof of stability of the adaptive FI-FV method is equivalent to the proof carried out for the adaptive FIT. The linear or spline interpolation operator can be applied, allowing for the evaluation of the continuous energy density $\overline{w}(\vec{r})$ (4.42) at every point $\vec{r} \in \Omega_\mathcal{G}$. From this point the proof follows the argumentation presented in the preceding section.

4.4 Time-Adaptive Grid Refinement for the Discontinuous Galerkin Method

Adaptive grid refinement for Finite Element Methods like the Discontinuous GALERKIN Method exhibits a basic difference from the FIT or FVM. The possibilities for adapting the local resolution in a region, i.e., the number of degrees of freedom, are twofold. Besides modifying the grid step size, the local approximation order $P(i)$ can be adjusted in an element-wise fashion. The former corresponds to the presented adaptation techniques for the FIT and the FVM. Within FE terminology, it is referred to as *h-refinement* while the latter is called *p-refinement*. In addition, both refinement techniques can be combined, yielding an *hp-adaptive* method.

In this section, first, a computationally efficient method for adapting the FE space along the h- and p-dimension is introduced. This is followed by a consideration of the accuracy and stability of the adaptation techniques. With these prerequisites at hand, the subject of an automated hp-adaptation of the computational grid is addressed in the last subsection.

4.4.1 Projection Based h-Adaptation

An adaptation of the local grid step size within a FE method demands for a modification of the local FES. In contrast to the interpolation based determination of new discrete field values for the FIT and FVM, a projection of the approximate solution from the current to the adapted FES has to be performed. In equation (3.107) on page 50 the basis functions were defined as scaled LEGENDRE functions within the interval I of width Δx and center x_0. In order to define a basis on the left and right hand halves of this interval the basis functions φ_l^l and φ_r^r are introduced according to:

$$\varphi_l^l(x) := \sqrt{2l+1}\, \mathcal{L}\left(\frac{4(x - (x_0 - \Delta x/4))}{\Delta x} \right) \tag{4.55}$$

$$\varphi_r^r(x) := \sqrt{2r+1}\, \mathcal{L}\left(\frac{4(x - (x_0 + \Delta x/4))}{\Delta x} \right) \tag{4.56}$$

with $l = 1..L$ and $r = 1..R$ (see fig. 4.9). They span the spaces $\mathcal{V}_l$ and $\mathcal{V}_r$ given by:

$$\mathcal{V}_l^L = \text{span}\{\varphi_l^l\} \tag{4.57}$$

$$\mathcal{V}_r^R = \text{span}\{\varphi_r^r\} \tag{4.58}$$

The approximation orders L and R in each subinterval do not have to be identical, neither are they required to be equal to the order P of the parent element. The direct sum of the refined spaces $\mathcal{V}_l$ and $\mathcal{V}_r$ is denoted by $\mathcal{V}^+$:

$$\mathcal{V}^+ = \mathcal{V}_l \oplus \mathcal{V}_r \tag{4.59}$$

In the following, the projection operator (3.97) can be applied in order to project the current approximation, given in an individual element i, to the FES defined on the h-refined or h-coarsened element. Due to the tensor product basis, the projections can be performed along each coordinate individually.

For the case of h-refinement the projections to the left and right hand subelement read:

$$(e_i)_l^l = \frac{\langle \varphi_l^l(x), \bar{E}_i(x) \rangle}{\langle \varphi_l^l(x), \varphi_l^l(x) \rangle} \quad \text{and} \quad (e_i)_r^r = \frac{\langle \varphi_r^r(x), \bar{E}_i(x) \rangle}{\langle \varphi_r^r(x), \varphi_r^r(x) \rangle} \tag{4.60}$$

$$(h_i)_l^l = \frac{\langle \varphi_l^l(x), \bar{H}_i(x) \rangle}{\langle \varphi_l^l(x), \varphi_l^l(x) \rangle} \quad \text{and} \quad (h_i)_r^r = \frac{\langle \varphi_r^r(x), \bar{H}_i(x) \rangle}{\langle \varphi_r^r(x), \varphi_r^r(x) \rangle} \tag{4.61}$$

Inserting the definitions of $\bar{E}$ and $\bar{H}$, (3.89) and (3.90), yields:

$$\begin{aligned} (e_i)_l^l &= \frac{\langle \varphi_l^l(x), \sum_p e_i^p \varphi^p(x) \rangle}{\langle \varphi_l^l(x), \varphi_l^l(x) \rangle} \\ &= \sum_p \frac{\langle \varphi_l^l(x), \varphi^p(x) \rangle}{\langle \varphi_l^l(x), \varphi_l^l(x) \rangle} e_i^p \\ &= \sum_p (\Pi_l^{l,p})^+ e_i^p \end{aligned} \tag{4.62}$$

with all other coefficients given accordingly. This defines the operators for performing the projections to the refined elements:

$$(\Pi_l^{l,p})^+ := \frac{\langle \varphi_l^l, \varphi^p \rangle}{\langle \varphi_l^l, \varphi_l^l \rangle} \tag{4.63}$$

$$(\Pi_r^{r,p})^+ := \frac{\langle \varphi_r^r, \varphi^p \rangle}{\langle \varphi_r^r, \varphi_r^r \rangle} \tag{4.64}$$

Since the definitions of the projectors are independent of the solution and involve only the basis functions φ_l^l, φ_r^r and φ^p, they can be analytically evaluated for all combinations of (l,p) and (r,p). The results are tabulated in the $(L \times P)$ and $(R \times P)$ matrices $\mathbf{\Pi}_l^+$ and $\mathbf{\Pi}_r^+$ (see fig. 4.9). The numerical DoF of the refined cells are subsequently obtained by the matrix-vector multiplications:

$$(\mathbf{e}_i)_l = \mathbf{\Pi}_l^+ \mathbf{e}_i \quad \text{and} \quad (\mathbf{e}_i)_r = \mathbf{\Pi}_r^+ \mathbf{e}_i \tag{4.65}$$

$$(\mathbf{h}_i)_l = \mathbf{\Pi}_l^+ \mathbf{h}_i \quad \text{and} \quad (\mathbf{h}_i)_r = \mathbf{\Pi}_r^+ \mathbf{h}_i \tag{4.66}$$

For the case of h-coarsening, the current approximation on the coarsened element is piecewise defined on the refined cells:

$$(\bar{E}_i(x))^+ = \begin{cases} (\bar{E}_i)_l(x) & \text{if } x \in (V_i)_l \\ (\bar{E}_i)_r(x) & \text{if } x \in (V_i)_r \end{cases} \tag{4.67}$$

and $(\bar{H}_i)^+$ given respectively. Using the property (4.59), this is rewritten as:

$$(\bar{E}_i(x))^+ = (\bar{E}_i)_l(x) + (\bar{E}_i)_r(x) \tag{4.68}$$

4.4. Time-Adaptive Grid Refinement for the Discontinuous GALERKIN Method

The current solution is projected to the FES of the coarsened element:

$$\begin{aligned}
e_i^p &= \frac{\left\langle \varphi^p(x), (\bar{E}_i(x))^+ \right\rangle}{\langle \varphi^p(x), \varphi^p(x) \rangle} \\
&= \frac{\left\langle \varphi^p(x), (\bar{E}_i)_l(x) \right\rangle}{\langle \varphi^p(x), \varphi^p(x) \rangle} + \frac{\left\langle \varphi^p(x), (\bar{E}_i)_r(x) \right\rangle}{\langle \varphi^p(x), \varphi^p(x) \rangle} \\
&= \frac{\left\langle \varphi^p(x), \sum_l (e_i)_l^l \varphi_l^l(x) \right\rangle}{\langle \varphi^p(x), \varphi^p(x) \rangle} + \frac{\left\langle \varphi^p(x), \sum_r (e_i)_r^r \varphi_r^r(x) \right\rangle}{\langle \varphi^p(x), \varphi^p(x) \rangle} \\
&= \sum_l \frac{\left\langle \varphi^p(x), \varphi_l^l(x) \right\rangle}{\langle \varphi^p(x), \varphi^p(x) \rangle} (e_i)_l^l + \sum_r \frac{\left\langle \varphi^p(x), \varphi_r^r(x) \right\rangle}{\langle \varphi^p(x), \varphi^p(x) \rangle} (e_i)_r^r \\
&= \sum_l (\Pi_l^{p,l})^- (e_i)_l^l + \sum_r (\Pi_r^{p,r})^- (e_i)_r^r
\end{aligned} \qquad (4.69)$$

with all other coefficients given accordingly. This defines the operators for performing the projection to the coarsened element:

$$(\Pi_l^{p,l})^- := \frac{\langle \varphi^p, \varphi_l^l \rangle}{\langle \varphi^p, \varphi^p \rangle} \qquad (4.70)$$

$$(\Pi_r^{p,r})^- := \frac{\langle \varphi^p, \varphi_r^r \rangle}{\langle \varphi^p, \varphi^p \rangle} \qquad (4.71)$$

Evaluating all combinations of (p,l) and (p,r) analytically gives rise to the $(P \times L)$ and $(P \times R)$ matrices $\mathbf{\Pi}_l^-$ and $\mathbf{\Pi}_r^-$. The DoF of the coarsened cell are obtained by:

$$\begin{aligned}
\mathbf{e}_i &= \mathbf{\Pi}_l^- (\mathbf{e}_i)_l + \mathbf{\Pi}_r^- (\mathbf{e}_i)_r \qquad &(4.72) \\
\mathbf{h}_i &= \mathbf{\Pi}_l^- (\mathbf{h}_i)_l + \mathbf{\Pi}_r^- (\mathbf{h}_i)_r \qquad &(4.73)
\end{aligned}$$

The projection matrices are listed in the appendix B. They are connected by the relations:

$$\mathbf{\Pi}_l^+ = 2 \left(\mathbf{\Pi}_l^- \right)^\mathrm{T} \quad \text{and} \quad \mathbf{\Pi}_r^+ = 2 \left(\mathbf{\Pi}_r^- \right)^\mathrm{T} \qquad (4.74)$$

The evaluation of a matrix-vector product is a computationally fast operation. The upper or lower triangular structure of the matrices, moreover, allows for an in-place calculation which avoids the usage of temporary memory. This provides a highly efficient framework for performing h-adaptations within in the DGM.

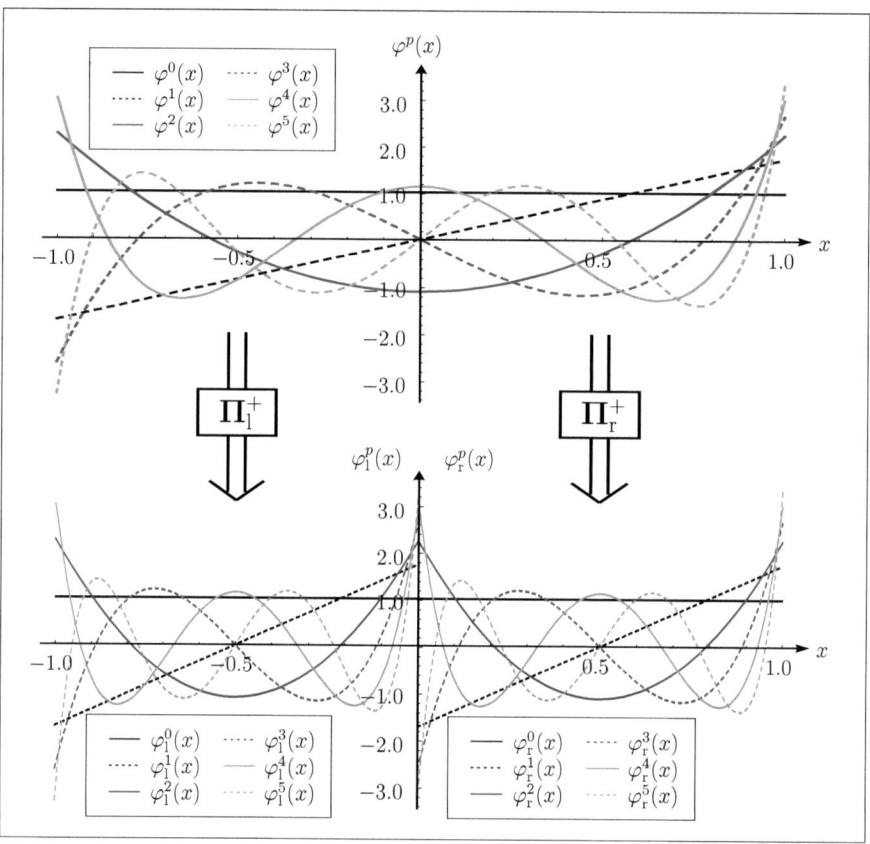

Figure 4.9: Basis functions and projection operation for h-refinement: The graphs show the basis functions for the coarse and refined grid and their support. The operators $\mathbf{\Pi}_l^+$ and $\mathbf{\Pi}_r^+$ project the solution to the refined elements.

4.4.2 Projection Based p-Adaptation

The orthogonality of the basis functions makes the adaptation of the local approximation order P a trivial task.

For the case of a p-enrichment, the local basis is expanded with the $(P+1)$-th order basis functions:
$$\mathcal{V}_i^{P+1} = \mathcal{V}_i^P \cup \varphi_i^{P+1} \tag{4.75}$$
The vectors of coefficients $\mathbf{e}_i$ and $\mathbf{h}_i$ are extended with the new coefficients $(e_u)_i^{P+1}$ and $(h_u)_i^{P+1}$ with $u \in \{x,y,z\}$. Due to the orthogonality, the coefficients $(e_u)_i^{1..P}$ and $(h_u)_i^{1..P}$ remain unaltered under a projection from $\mathcal{V}_i^P$ to $\mathcal{V}_i^{P+1}$. The new coefficients are initialized to zero.

For the case of a p-reduction, the P-th order basis functions are removed from the local basis:
$$\mathcal{V}_i^{P-1} = \mathcal{V}_i^P \setminus \varphi_i^P \tag{4.76}$$
Following the same argument, the coefficients $(e_u)_i^P$ and $(h_u)_i^P$ are deleted from the vectors of DoF while the coefficients $(e_u)_i^{1..P-1}$ and $(h_u)_i^{1..P-1}$ remain unaltered.

While p-enrichment with hierarchical basis functions for continuous FEM works in exactly the same way, an optimal p-reduction requires more efforts. Since the basis is not orthogonal, the projection to the reduced FES is not trivial and demands for an actual computation and adaptation of all remaining coefficients ([45] pp. 74).

Despite the fact that p-refinement is mathematically trivial for the given case of orthogonal basis functions, a good implementation in a computer program is a challenging task. Especially the extension and reduction of the local vectors of DoF pose a problem for an efficient memory management. This issue is discussed in the appendix C.

4.4.3 Properties of the Projection Based Adaptations

Optimality and Accuracy

Using the HILBERT *Projection Theorem* 3.1 (p. 48), it can immediately be concluded that the projection based refinement operations for performing h- and p-adaptation are optimal in the sense of the *Optimality Theorem* 3.2. Nevertheless, some complementary remarks can be added.

In the case of a pure h-refinement, i.e., the approximation orders L, R and P are identical, the approximations $\overline{E}_i$ and $\overline{H}_i$ can always be represented exactly on the refined elements:
$$(\overline{E}_i)_\mathrm{l} + (\overline{E}_i)_\mathrm{r} \equiv \overline{E}_i \quad \text{and} \quad (\overline{H}_i)_\mathrm{l} + (\overline{H}_i)_\mathrm{r} \equiv \overline{H}_i \tag{4.77}$$
as shown in figure 4.10a. This holds true because the space $\mathcal{V}^+$ (4.59) spanned by the basis functions of the refined elements is larger than the space $\mathcal{V}$ of the parent element:
$$\mathcal{V} \subset \mathcal{V}^+ \quad \text{for} \quad L = R = P \tag{4.78}$$
Consequently, during the inverse procedure of h-coarsening, the piecewise defined approximations of the electric field $(\overline{E}_i)^+$ (4.68) and the magnetic field $(\overline{H}_i)^+$ can, in general, not be represented exactly on the coarsened cell:
$$(\overline{E}_i)^+ \not\equiv \overline{E}_i \quad \text{and} \quad (\overline{H}_i)^+ \not\equiv \overline{H}_i \tag{4.79}$$

Due to the discontinuous nature of these piecewise defined functions, this holds true for any finite approximation order P of the coarsened cell. Figure 4.10b illustrates this.

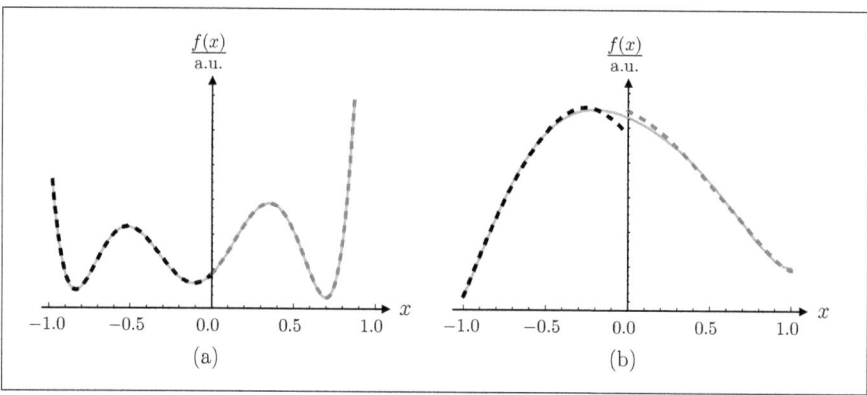

Figure 4.10: Projection based grid refinement and coarsening: In (a) the projection of a function to the left and right hand side refined element is shown. If $L = R = P$, this projection is always exact because the space spanned by the basis functions of the refined cells is larger. In consequence, the projection to the coarsened cell shown in (b) can, in general, not be exact. This statement remains true for any finite P because of the discontinuity at the interface.

Stability

In order to prove stability for the h- and p-adaptation techniques introduced above, first, the energy norm (3.123) defined on page 52 is examined more closely. Since the approximations of the electric and magnetic field are cell-wise continuous functions, the continuous definition of the energy density w (2.20) can be adapted to the DGM:

$$\overline{w}_i(\vec{r}) = \frac{1}{2}\left(\overline{E}_i(\vec{r}) \cdot \epsilon_i \overline{E}_i(\vec{r}) + \overline{H}_i(\vec{r}) \cdot \mu_i \overline{H}_i(\vec{r})\right) \quad (4.80)$$

The summation of the integrated energy density within every cell yields a norm for the discrete electromagnetic energy within the computational domain.

Evaluating the discrete energy of the cell i reads:

$$\begin{aligned}
\mathcal{W}_i &= \int_{\mathcal{G}_i} \overline{w}_i(\vec{r}) \, dV \\
&= \frac{1}{2} \int_{\mathcal{G}_i} \left(\overline{E}_i(\vec{r}) \cdot \epsilon_i \overline{E}_i(\vec{r}) + \overline{H}_i(\vec{r}) \cdot \mu_i \overline{H}_i(\vec{r})\right) dV \\
&= \frac{1}{2} \int_{\mathcal{G}_i} \left(\mathbf{e}_i^T \mathbf{\Phi} \cdot \epsilon_i \mathbf{e}_i^T \mathbf{\Phi} + \mathbf{h}_i^T \mathbf{\Phi} \cdot \mu_i \mathbf{h}_i^T \mathbf{\Phi}\right) dV \\
&= \frac{1}{2} V_i \left(\epsilon_i \mathbf{e}_i^T \mathbf{e}_i + \mu_i \mathbf{h}_i^T \mathbf{h}_i\right) \\
&= \frac{1}{2} V_i \left(\epsilon_i \|\mathbf{e}_i\|_2^2 + \mu_i \|\mathbf{h}_i\|_2^2\right)
\end{aligned} \quad (4.81)$$

4.4. Time-Adaptive Grid Refinement for the Discontinuous GALERKIN Method

where the vector notations (3.92) and (3.108) and the orthogonality property (3.110) were used. Using the definition of the elements of the mass matrix (3.111) yields the norm (3.123).

The tensor product basis allows for the individual consideration of each coordinate direction. It is, therefore, sufficient to show that the energy (4.81) is non-increasing during any adaptation in a one-dimensional domain. This domain may consist of only one cell.

h-Adaptation: For the following discussion of stability it is assumed that the maximum approximation orders L, R and P are identical. It is clarified later, that this does not pose a restriction to the general validity of the results.

In the case of h-refinement, the operators Π_l^+ and Π_r^+ project from the space $\mathcal{V}$ to the larger space $\mathcal{V}^+$. Following the argument of the preceding subsection on optimality and accuracy, any function defined in the space $\mathcal{V}$ is exactly represented in $\mathcal{V}^+$. The conservation of the discrete energy is a direct consequence[3]:

$$\begin{aligned}
\mathcal{W}_i &= (\mathcal{W}_i)_l + (\mathcal{W}_i)_r \\
\Longleftrightarrow \int_{-\Delta x/2}^{\Delta x/2} \overline{w}_i(x)\,\mathrm{d}x &= \int_{-\Delta x/2}^{0} (\overline{w}_i)_l(x)\,\mathrm{d}x + \int_{0}^{\Delta x/2} (\overline{w}_i)_r(x)\,\mathrm{d}x \\
\Longleftrightarrow \tfrac{\Delta x}{2}\left(\epsilon_i\|\mathbf{e}_i\|_2^2 + \mu_i\|\mathbf{h}_i\|_2^2\right) &= \tfrac{\Delta x}{4}\left(\epsilon_i\|(\mathbf{e}_i)_l\|_2^2 + \mu_i\|(\mathbf{h}_i)_l\|_2^2 + \epsilon_i\|(\mathbf{e}_i)_r\|_2^2 + \mu_i\|(\mathbf{h}_i)_r\|_2^2\right) \\
\Longleftrightarrow 2\left(\epsilon_i\|\mathbf{e}_i\|_2^2 + \mu_i\|\mathbf{h}_i\|_2^2\right) &= \epsilon_i(\|(\mathbf{e}_i)_l\|_2^2 + \|(\mathbf{e}_i)_r\|_2^2) + \mu_i(\|(\mathbf{h}_i)_l\|_2^2 + \|(\mathbf{h}_i)_r\|_2^2)
\end{aligned} \qquad (4.82)$$

The coarse grid function plotted in figure 4.10a has a maximum order of $P = 6$ with all coefficients equal to one. The coefficients of the functions on the refined grid and the square values of their 2-norms are given in the table 4.1. If the vector $\mathbf{c}$ is considered to be either the vector of coefficients of the electric field $\mathbf{e}$ or the magnetic field $\mathbf{h}$, the equation (4.82) is fulfilled.

	c_0	c_1	c_2	c_3	c_4	c_5	c_6	$\|\mathbf{c}\|_2^2$
$f(x)$	1.0000	1.0000	1.0000	1.0000	1.0000	1.0000	1.0000	7.0000
$f(x)_l$	0.2574	-0.1355	-0.2606	-0.1446	0.4276	-0.1563	0.0156	0.3808
$f(x)_r$	1.7426	1.5631	1.6819	2.0399	1.0494	0.2181	0.0156	13.6192

Table 4.1: Coarse and fine grid coefficients of the function plotted in figure 4.10a

The h-coarsening operators Π_l^- and Π_r^- project a function from the space $\mathcal{V}^+$ to the smaller space $\mathcal{V}$. Since $\mathcal{V}$ is a subspace of $\mathcal{V}^+$, it is immediately concluded that, in general, energy is lost during the coarsening process. The discrete energy can only be preserved if the union of the left and right hand functions is an element of the smaller space $\mathcal{V}$. Starting with the coefficients of the refined cells, given in the table 4.1, the coarse grid coefficients are exactly recovered and the discrete energy is preserved.

In addition, it can be shown from algebraic properties of the projection matrices, that the discrete energy for arbitrary fine grid coefficients is always non-increasing during h-coarsening. First, the $(P \times 2P)$ projection matrix Π^- is defined as:

$$\Pi^- := \begin{pmatrix} \Pi_l^- & \Pi_r^- \end{pmatrix} \qquad (4.83)$$

---

[3] Identical material properties are assumed for the parent cell and the two refined cells.

The coefficients of the two refined cells are gathered in one vector $(\mathbf{c}_i)^+$:

$$(\mathbf{c}_i)^+ := \begin{pmatrix} (\mathbf{c}_i)_l \\ (\mathbf{c}_i)_r \end{pmatrix} \tag{4.84}$$

Then, the coefficients of the coarsened cell are given by:

$$\mathbf{c}_i = \mathbf{\Pi}^-(\mathbf{c}_i)^+ \tag{4.85}$$

which is an equivalent notation of the equations (4.72) and (4.73). Using (4.81) and (4.82), the following must hold true in order to guarantee a non-increasing discrete energy:

$$\begin{aligned}
& 2\|\mathbf{c}_i\|_2^2 \overset{!}{\leq} \|(\mathbf{c}_i)^+\|_2^2 \\
\Longleftrightarrow\quad & 2\,\mathbf{c}_i^T \mathbf{c}_i \overset{!}{\leq} ((\mathbf{c}_i)^+)^T(\mathbf{c}_i)^+ \\
\Longleftrightarrow\quad & 2\,(\mathbf{\Pi}^-(\mathbf{c}_i)^+)^T \mathbf{\Pi}^-(\mathbf{c}_i)^+ \overset{!}{\leq} ((\mathbf{c}_i)^+)^T(\mathbf{c}_i)^+ \\
\Longleftrightarrow\quad & \frac{((\mathbf{c}_i)^+)^T (\mathbf{\Pi}^-)^T \mathbf{\Pi}^-(\mathbf{c}_i)^+}{((\mathbf{c}_i)^+)^T(\mathbf{c}_i)^+} \overset{!}{\leq} \frac{1}{2}
\end{aligned} \tag{4.86}$$

In order to fulfill this, it is sufficient to demand:

$$\|\left(\mathbf{\Pi}^-\right)^T \mathbf{\Pi}^-\|_2^2 \leq \frac{1}{2} \tag{4.87}$$

or equivalently:

$$\max\left\{\mathrm{eig}\left(\left(\mathbf{\Pi}^-\right)^T \mathbf{\Pi}^-\right)\right\} \leq \frac{1}{2} \tag{4.88}$$

It can be verified that the matrix $\left((\mathbf{\Pi}^-)^T \mathbf{\Pi}^-\right)$ has the P-times degenerated eigenvalues 0.5 and 0, which completes the proof.

The coefficients for the example shown in figure 4.10b are given in the table 4.2. The sum of the energy norms for the left and right hand refined cells yields 1.9003 while twice the norm of the coarsened cell evaluates to 1.8996. Thus, energy was lost during coarsening.

	c_0	c_1	c_2	c_3	$\|\mathbf{c}\|_2^2$
$f(x)$	0.9500	-0.0433	-0.2073	0.0498	0.9498
$f(x)_l$	1.0000	0.2000	-0.1000	-0.0100	1.0501
$f(x)_r$	0.9000	-0.2000	-0.0100	0.0100	0.8502

Table 4.2: Coarse and fine grid coefficients of the function plotted in figure 4.10b

p-Adaptation: The proofs of stability for the introduced p-adaptations are straightforward. In the case of a p-enrichment, the vectors of DoF are extended by the coefficients for the $(P+1)$-th order basis functions. Since these coefficients are initialized to zero, it holds true:

$$\|(\mathbf{e})^P\|_2^2 = \|(\mathbf{e})^P\|_2^2 + 0 = \|(\mathbf{e}_{0..P})^{P+1}\|_2^2 + (e^{P+1})^2 = \|(\mathbf{e})^{P+1}\|_2^2 \tag{4.89}$$

for the electric field and the magnetic field respectively. The discrete energy is exactly conserved.

In the case of a p-reduction, the coefficients assigned to the highest order basis functions are removed from the vectors of DoF. Consequently, it holds true:

$$\|(\mathbf{e})^P\|_2^2 = \|(\mathbf{e}_{0..P-1}))^P\|_2^2 + (e^P)^2 \leq \|(\mathbf{e}_{0..P-1})^P\|_2^2 = \|(\mathbf{e})^{P-1}\|_2^2 \qquad (4.90)$$

and the discrete energy is at most preserved or otherwise reduced.

In the discussion of stability for h-adaptations it was assumed that the maximum approximation orders L, R and P are identical. After showing that the p-adaptation is stable, it can be concluded that this assumption does not restrict the validity of the results obtained there. If the orders L and R are not equal, a p-adaptation can be applied first in order to match them. This reduces the problem to a sequential execution of stable p- and h-adaptations.

Finally, some remarks can be made about mixed h- and p-adaptations. If an h-coarsening goes along with a p-enrichment, it is beneficial to employ projection matrices Π_l^- and Π_r^- of the size $(P+1 \times P)$. In this case the loss of accuracy is diminished in comparison to a projection to the space $\mathcal{V}^P$ and a subsequent extension of the vectors of DoF with zeros. On the contrary, projection matrices Π_l^+ and Π_r^+ of the size $(P-1 \times P)$ can be employed, if an h-refinement and a p-reduction should be carried out simultaneously.

4.4.4 Automated hp-Adaptivity

In the preceding sections, techniques for the adaptation of the computational grid within the framework of the FIT, FVM and DGM have been introduced, and their stability was proven. However, the topic of refinement indicators has not been addressed. For the case of PIC simulations this indicator is trivially prescribed. The grid has to be refined in a neighborhood of the particles in order to be able to resolve the high-frequency space charge fields. With increased distance from the particles the grid can be gradually coarsened. This works particularly well if the particles are bunched, which is true for the applications covered in the chapter 5.

Nevertheless, the topic of automated grid adaptation, based on the electromagnetic field distribution is attractive. This is especially true for the DGM because of its twofold refinement mechanisms, h- and p-refinement. The remainder of this section is concerned with the development of an algorithm for performing automated hp-adaptivity for the DGM. The subject of automated grid adaptation is closely linked to error estimation techniques. A review of error estimation in the context of the DGM is found in [35, 67]. However, the error estimator does only indicate cells that demand for an adaptation of the local resolution. It does not give any immediate information about which kind of adaptation is best suited. It requires a second criterion to decide upon h- or p-adaptation, which is based on the local smoothness of the field distribution.

The investigation and development of hp-adaptive methods (FEM or DGM) is an active research area and equally popular among mathematicians, physicists and engineers even though from a different point of view. Nevertheless, a review of the available literature reveals that there is no *optimal* solution to the problem yet. Existing implementations are usually limited to problems in the two-dimensional space (see e.g. [2, 37, 100]). Moreover, the term of an hp-adaptive method is also used in conjunction with elliptic problems. There, it refers to the determination of the so-called *minimal hp-grid* [89]. In that context, hp-refinement is applied in order to obtain a solution that fulfills some accuracy requirements with a minimal number of DoF.

The automated grid adaptation method presented here, does not aim for the smallest number of DoF possible but for robustness. If necessary, a larger number of DoF is always accepted in exchange for a more robust scheme. In addition, the practicability of the scheme is a major concern. A fully adaptive scheme is only of practical use, if the computational time spent for the grid adaptation is outweighed by the savings in computing the actual solution. The scheme is developed in the one-dimensional space for a wave propagation along the x-coordinate. Due to the tensor product character of the employed basis, an extension to the three-dimensional space is straightforward.

Convergence of h- and p-Refinement

The studies on the convergence order of the DGM, presented in the sections 3.4.2 for the semidiscrete formulation and 3.6 for the fully discretized equations, indicate that it is advisable to choose the approximation order P as large as possible. However, the accuracy orders established in (3.180) are only valid in the asymptotic limit of $\Delta x \to 0$. Moreover, all analyses assume smooth solutions. For such solutions the discretization error in the L^2-norm converges according to [60]:

Theorem 4.1 (Convergence order). *Let $U(x,t)$ be a smooth solution of the wave equation (2.32) and $\overline{U}(x,t)$ the solution of the wave equation obtained by the DG formulation. Then the spatial discretization error $\mathcal{E}$ converges according to:*

$$\mathcal{E} := \|U - \overline{U}\|_2 \leq C(\Delta x)^P \tag{4.91}$$

with C a constant, Δx the element size, and P the maximum polynomial order of the basis.

Figure 4.11 illustrates the convergence of the discretization error in the L^2-norm for a GAUSSIAN and a trapezoidal function with regard to the grid step size. The convergence order strongly depends on the smoothness of the function. While the error for the GAUSSIAN distribution converges to zero with the theoretically expected order, it breaks down for the case of the non-smooth trapezoidal distribution. This is a well-known behavior and reported in, e.g., [45, 74]. Consequently, it is not generally preferable to perform p-refinement. Nevertheless, the h-refined neighborhood of a non-smooth region should be as small as possible in order to fully benefit from the superior convergence of high-order approximations.

Error Estimation

In [34] a relation between the size of the jumps on the element boundaries:

$$[\![\overline{E}]\!] = \overline{E}^- - \overline{E}^+ \quad \text{and} \quad [\![\overline{H}]\!] = \overline{H}^- - \overline{H}^+ \tag{4.92}$$

and the residual within in the respective element is derived. This is a valuable tool for the development of an adaptive DG method because the evaluation of these jumps is a computationally inexpensive operation. Moreover, it can be incorporated with the calculation of the flux terms which renders the extra costs negligible. The derivation in [34] is conducted using upwind fluxes. A similar expression can be found for central fluxes. Its derivation is outlined in the following.

4.4. Time-Adaptive Grid Refinement for the Discontinuous GALERKIN Method

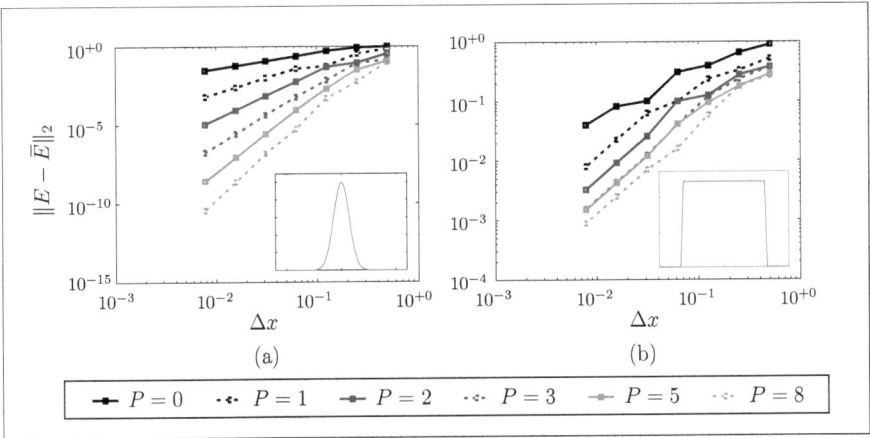

Figure 4.11: Dependence of the convergence rate on the smoothness of the function: L^2-norm of the discretization error for a GAUSSIAN (a) and a trapezoidal (b) function vs. the grid step size. The convergence rate obviously depends on the smoothness of the function. While the convergence of the GAUSSIAN distribution agrees with the expected order for all $P = 0..8$ (see the *Convergence Order Theorem 4.1*), no substantial benefit can be achieved from an increased approximation order for the trapezoidal distribution.

The DG formulation of FARADAY's law for a single element in one-dimensional space reads:

$$\int_{\mathcal{G}_i} \varphi^q \, \mathrm{d}_t \overline{H}_i \mathrm{d}x + (\overline{E}_i^* \varphi^q)^+ - (\overline{E}_i^* \varphi^q)^- = \int_{\mathcal{G}_i} \overline{E}_i \partial_x \varphi^q \mathrm{d}x \qquad \forall q = 1..P \tag{4.93}$$

Performing an integration by parts of the right hand side and rearranging the terms yields:

$$\int_{\mathcal{G}_i} \varphi^q \left(\mathrm{d}_t \mu \overline{H}_i + \partial_x \overline{E}_i \right) \mathrm{d}x = \left[\overline{E}_i \varphi^q \right]_{\partial \mathcal{G}_i} - (\overline{E}_i^* \varphi^q)^+ + (\overline{E}_i^* \varphi^q)^- \qquad \forall q = 1..P \tag{4.94}$$

which allows for the definition of the residual R:

$$R_i := \mathrm{d}_t \mu \overline{H}_i + \partial_x \overline{E}_i \tag{4.95}$$

Inserting the flux definition (3.116) and setting $\varphi = 1$ leads to the final result:

$$\int_{\mathcal{G}_i} R_i \, \mathrm{d}x = \frac{[\![\overline{E}_i]\!]^+}{2} + \frac{[\![\overline{E}_i]\!]^-}{2} \tag{4.96}$$

which establishes the link between the jump sizes and the residual. The above equation can be employed as a local error estimate.

An even simpler estimator can be constructed if the orthogonality of the basis is exploited. The *Convergence Order Theorem* 4.1 states that, at least in smooth regions, the order of the discretization error is $P + 1$:

$$\mathcal{E}_{\mathcal{G}_i} = \mathcal{O}(P+1) \tag{4.97}$$

The leading error term of the DG solution for the electric field is given by:

$$\|(E-\bar{E})_{\mathcal{G}_i}\|_2 \simeq \left(\int_{\mathcal{G}_i}\left(e_i^{P+1}\varphi^{P+1}\right)^2\mathrm{d}x\right)^{1/2} = \sqrt{\Delta x}|e_i^{P+1}| \qquad (4.98)$$

The coefficient of the next higher order, $P+1$, would be an estimate of the actual error in the element while the previous estimator refers to its residual. Since this coefficient is unknown, the absolute value of the coefficient of the presently highest order P is employed as an error estimator:

$$\|(E-\bar{E})_{\mathcal{G}_i}\|_2 \approx \sqrt{\Delta x}|e_i^P| \qquad (4.99)$$

In smooth regions and for sufficiently good approximations, this technique has shown a reasonable performance.

Both estimators were thoroughly investigated throughout a master thesis. See [122] for a summary of the results.

Smoothness Estimation

In the literature some techniques for smoothness estimation can be found. Knowing about the theoretical convergence rate of the DGM for smooth solutions, the local regularity of the true solution can be estimated from a comparison of solutions obtained with different approximation orders [4]. A related idea is presented in [100]. There, the amplitude of the jumps of the numerical solution across inter-element boundaries is checked for different approximation orders. For smooth solutions this property exhibits a super-convergence.

Both techniques require solutions obtained with diverse approximation orders. A solution obtained with a basis of maximum order P does inherently include all lower order solutions as a consequence of the orthogonality property. Nevertheless, in order to apply the methods, at least piecewise quadratic basis functions have to be used. This poses a rather stringent limit to the lowest number of applicable DoF per cell.

Hence, other possibly heuristic estimators were sought. A simple heuristic estimator is obtained by comparing the difference of the upper and lower interface value of an element and its neighbors. The difference for the electric field is given by:

$$\widehat{G}_i = \frac{\bar{E}^*_{i+1/2} - \bar{E}^*_{i-1/2}}{\Delta x_i} \qquad (4.100)$$

This value can be interpreted as a gradient-like property. They are gathered in a vector $\widehat{\mathbf{G}}$ and sorted by the absolute value. Cells showing a small gradient are not considered during the subsequent procedure. A high gradient alone is, nevertheless, not a sign for a low regularity of the solution. It is rather meaningful to compare the respective gradients of neighboring cells. This is expressed in the characteristic value G:

$$G_i = \max\left\{\left|\frac{\widehat{G}_i + \widehat{G}_{i-1}}{\min\{\widehat{G}_i, \widehat{G}_{i-1}\}}\right|, \left|\frac{\widehat{G}_i + \widehat{G}_{i+1}}{\min\{\widehat{G}_i, \widehat{G}_{i+1}\}}\right|\right\} \qquad (4.101)$$

Adding up the differences of neighboring cells leads to a partial cancellation of oppositely signed values which is commonly a sign for numerical high-frequency noise. Areas exhibiting a high

4.4. Time-Adaptive Grid Refinement for the Discontinuous GALERKIN Method

level of numerical noise are not supposed to be further refined. Dividing by the smaller of the two differences increases the characteristic value. The increase is even more significant if the difference between them is large. A large characteristic value is regarded as a sign of low regularity. The procedure is illustrated in the figure 4.12 for an exemplary setup.

Of course, there are two points which require an active interference and some *trial-and-error* setting of parameters. First, a limit has to be set in order to choose which cells are not considered because the gradient $\hat{G}$ is too small. Second, a limit of the characteristic value G has to be set. All cells exceeding this limit, and their immediate neighbors, are considered to reside in a region of low regularity and will be h-refined.

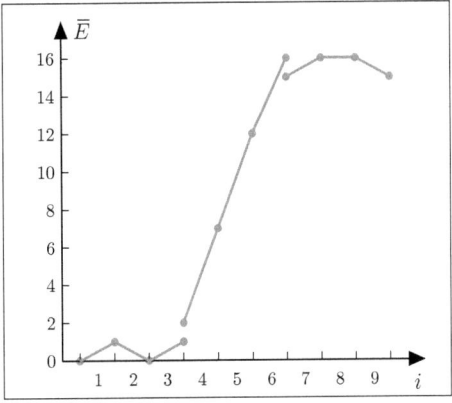

Figure 4.12: Exemplary distribution of the approximated electric field using piecewise linear basis functions. The table lists $\hat{G}$ and G:

i	1	2	3	4	5	6	7	8	9
$\hat{G}_i$	1	-1	1	5	5	4	1	0	-1
G_i	x	x	x	6	9/4	5	x	x	x

An 'x' indicates that the characteristic value is not evaluated because the respective gradient is too small. The largest characteristic values are computed for the cells 4 and 6 which are the regions of lowest regularity.

A second technique for estimating smoothness is based on the comparison of the highest order coefficient of adjacent cells. Following the *Convergence Order Theorem* 4.1, the error of the approximated solution tends towards zero in P-th order if the true solution is smooth. The motivation for this indicator is the following: In smooth regions, the true solution can be approximated well and the highest order coefficient is small. In non-smooth regions, the convergence order breaks down. The true solution cannot be approximated well anymore, and the highest order coefficient is large. A comparison of this coefficient in neighboring cells, thus, reveals regions of low regularity. The indicator is computed as follows:

$$C_i^1 = \log \left| \frac{e_i^P}{e_{i-1}^P} \right| \quad \text{with} \quad P = \max\{P_i, P_{i-1}\} \quad (4.102)$$

$$C_i^2 = \log \left| \frac{e_i^P}{e_{i+1}^P} \right| \quad \text{with} \quad P = \max\{P_i, P_{i+1}\} \quad (4.103)$$

$$C_i = \max\{|C_i^1|, |C_i^2|\} \quad (4.104)$$

The first indicator, G, is purely heuristic while the second indicator, C, is motivated from properties of the DG scheme. It depends on the problem under consideration which indicator performs better [122].

Adaptation Strategy

The introduction of error and smoothness estimators, in combination with the presented h- and p-adaptation techniques complete the prerequisites for implementing an autonomously working hp-adaptive scheme. It remains to define the adaptation strategy involving these building blocks. Again, detailed results of the various options are found in [122] while here only the most successful strategy is outlined.

This strategy is an extension of the approach presented in [99]. It maps a designated refinement level to every cell according to the distribution of the cell-wise estimated errors. This mapping is controlled by a progressivity parameter. It requires the definition of an identifier describing the resolution of a cell, η_i, which is defined as a function of the local approximation order P_i and the level of h-refinement L_i. A possible choice ([122] p. 47) is:

$$\eta_i := L_i \cdot P_i \quad (4.105)$$

After every N-th time step, the error estimator is evaluated yielding the estimate $\bar{\mathcal{E}}_i$ for every cell. All cells are assigned a designated resolution η_i^* specified by:

$$\eta_i^* := \max\{\hat{\eta} - j, 0\} \quad (4.106)$$

with $\hat{\eta}$ a preset maximum resolution and j such that:

$$\frac{\bar{\mathcal{E}}_{\max}}{d^{j+1}} \leq \bar{\mathcal{E}}_i < \frac{\bar{\mathcal{E}}_{\max}}{d^j} \quad \text{and} \quad \bar{\mathcal{E}}_{\max} = \max_i\{\bar{\mathcal{E}}_i\} \quad (4.107)$$

The parameter d controls the progressivity of the mapping.

The scheme is best illustrated by an example. If d is set to two, all cells with $\bar{\mathcal{E}}_i \geq \bar{\mathcal{E}}_{\max}/2$ are assigned the maximum resolution $\eta_i^* = \hat{\eta}$. The next error interval comprises all cells with $\bar{\mathcal{E}}_{\max}/4 \leq \bar{\mathcal{E}}_i < \bar{\mathcal{E}}_{\max}/2$. The designated resolution assigned to these cells is $\eta_i^* = \hat{\eta} - 1$. The procedure continues until all cells are assigned a designated level. Cells with an estimated error of $\bar{\mathcal{E}}_i < \bar{\mathcal{E}}_{\max}/2^{\hat{\eta}}$ are assigned the the lowest resolution of $\eta_i^* = 0$. A comparison of the actual resolution η_i with the designated level η_i^* identifies the cells which require an adaptation. See figure 4.13 below for a graphical illustration.

4.4. Time-Adaptive Grid Refinement for the Discontinuous GALERKIN Method 119

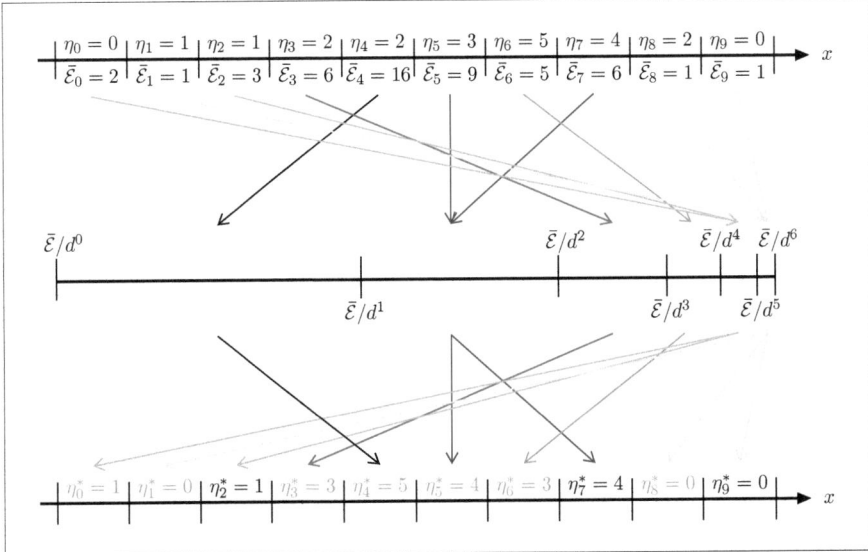

Figure 4.13: Illustration of the strategy for resolution adaptation: The error estimator is evaluated for all elements yielding the values $\bar{\mathcal{E}}_i$. Those are mapped to the error intervals given by equation (4.107) with the progressivity parameter d. In the figure, the maximum error is denoted by $\bar{\mathcal{E}}$. A cell-wise evaluation of (4.106), with a maximum level of $\hat{\eta} = 5$ in this example, determines the designated resolution η_i^* which is compared with the current resolution η_i. If the levels differ, the cell is flagged for an adaptation. The designated resolution of the respective cells is marked in gray.

Next, it has to be decided upon an h- or p-adaptation. Generally, p-refinement is preferred for its superior convergence rate, unless a region of low regularity is indicated. In case of an h-refinement, the approximation order of the refined elements are chosen such that $P_i^{\mathrm{I}} = P_i^{\mathrm{II}} = \lfloor P_i/2 \rfloor^4$. This refinement is called *competitive* since it leads to an equal or similar number of DoF of the parent and child elements.

If the local resolution can be reduced, the current solution is projected to the h- and p-reduced FES. In order to obtain competitive adaptations, the approximation order of the h-reduced element is allowed to be up to $P_i = P_i^{\mathrm{I}} + P_i^{\mathrm{II}}$. The evaluation of the error estimate for all cases yields a criterion for choosing the kind of reduction.

In order to enhance the robustness, this strategy is supplemented by the following constraints:

- If the relative electromagnetic energy stored in an element is small, it is not considered for refinement, regardless of any indicator. In case the element is refined already, its resolution is decreased to the minimum.

---
[4] In this case the discrete energy is not strictly conserved during h-refinement because this operation can be expressed as a sequence of an energy conserving pure h-refinement and a p-reduction which generally reduces the stored energy.

- The h- and p-refinement levels of neighboring elements are not allowed to differ by more than one.

 - If a low regularity of the solution is identified, the approximation order of the current element and its neighbors is decreased while the level of h-refinement is increased.

 - A reduction of the local resolution is only performed if the estimated error does not increase by more than a factor of two.

The enforcement of these constraints leads to a more conservative scheme in the sense that usually a larger number of DoF is used. Since numerical experiments have shown the effectiveness of these modifications with regard to the robustness of the complete adaptation algorithm, this is accepted.

In the preceding sections and chapters, attention was paid to a reasonably profound mathematical presentation. The automated hp-adaptation scheme presented on the last pages, in contrast, is of a rather heuristic nature. This is especially true for the smoothness estimators and the adaptation strategy. Mathematically more substantiated schemes usually require solutions on multiple grids [45, 117]. This, however, is in disagreement with the goal of developing an adaptive scheme of practical usability. More details and illustrative flowcharts of the presented algorithm are found in [122].

5. TESTS AND APPLICATIONS

The discretization methods and grid adaptation techniques, introduced in the preceding chapters, have been implemented in a computer program. The development is partially documented throughout the publications [106–109]. In the following, a summary of further implementation details is given. The complete technical documentation is found in [125].

The program is written in the C programming language. All algorithmic parts are parallelized using the *OpenMP Application Program Interface (API)* for multi-platform shared-memory parallel programming [93] in order to take advantage of the presently widespread multi-processor architecture. The program is controlled using an input file which is parsed during program initialization. An exemplary input file is found in the appendix D. Grid and field data output is generated in the *Visualization Toolkit* (VTK) data file format [127]. The evolution of the particle bunch in terms of moments of the distribution during the simulation is stored in a single multi-column file. The memory management system is separated from the main program and linked to it as a *dynamic link library* (DLL). A more detailed description of the memory system is found in the appendix C.

In the following, the results of a sanity check of the implemented algorithms are presented. In 5.2, the numerical methods are benchmarked. The simulation of a non-relativistic bunch in a metallic tube is carried out applying all methods. For this example an analytical solution is available. In the introduction, the FLASH injector was shown as an example, which motivates the application of time-adaptive grid refinement. Results obtained from simulations of this injector are presented in 5.3.

5.1 Sanity Check

In software development, the sanity check is a brief run-through of the functionality of a computer program in order to assure that the system or methodology works as expected. In order to present instructive figures, the results are displayed in the form of one-dimensional plots along a path or as two-dimensional graphs in cut view. However, the tests were carried using the above described program for the solution of problems in three-dimensional space.

5.1.1 Grid Adaptation

In order to check the functionality of the grid topology adaptation, a base grid consisting of $4 \times 4 \times 25$ cells is set up. A fictional refinement region of a third of the total length is defined and moved along the z-coordinate. A refinement level of two is chosen. Snapshots of the resulting grid topologies are shown in the figure 5.1. A test of the adaptive material discretization routine yielded the material discretizations shown in figure 3.7b.

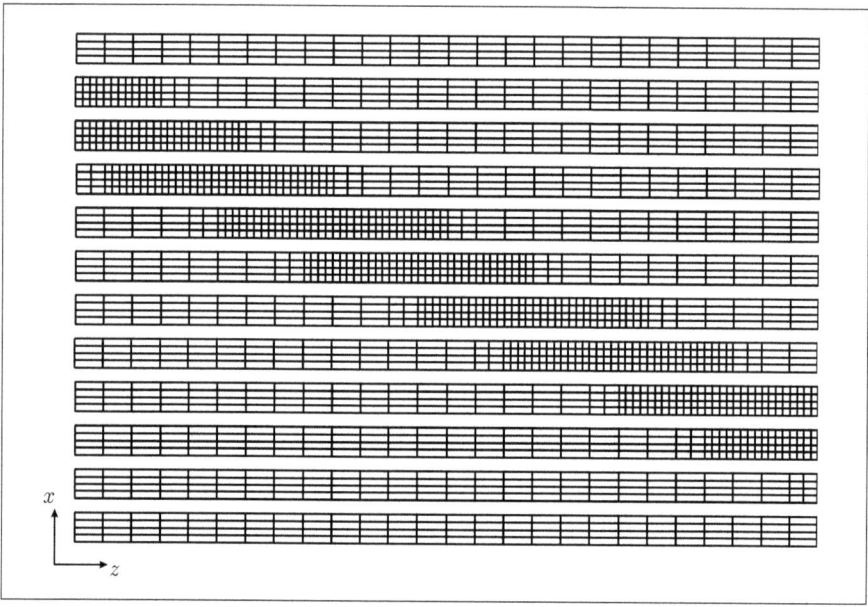

Figure 5.1: Verification of the grid adaptation functionality: Series of snapshots of the computational grid in cut view.

5.1.2 Interpolation Techniques

In order to verify the interpolation procedures a computational domain consisting of $1 \times 1 \times 20$ equally sized cells is set up. All field components are initialized using a polynomial function of first, second, third and fourth order (fig. 5.2a) followed by a single refinement step in all cells. For the determination of the field values the linear, the C-spline and the AKIMA spline interpolations are used. The interpolated values are compared with the analytical ones. For the linear distribution, no interpolation errors occur. The interpolation errors of the $\overline{e}_x$ voltages for all other cases are shown in the figures 5.2 b-d.

The C-spline interpolation shows large errors towards the lower and upper end of the domain. This results from the constraint of a vanishing second derivative of the natural cubic spline at interval boundaries (see sec. 4.2.1 p. 93). Otherwise, the errors agree with the expectations. The linear interpolation is erroneous for any non-linear function. For the quadratic case (b) the error of the AKIMA interpolation is zero to machine precision while the error of the C-spline interpolation is very small ($\approx 10^{-9}$) in the middle of the domain but non-zero because of the errors introduced at the domain boundaries and the global character of the spline. For higher order distributions (c,d) the errors of the spline interpolations are small but non-zero.

5.1. Sanity Check

It is concluded that all interpolations work as expected with the AKIMA spline interpolation performing best.

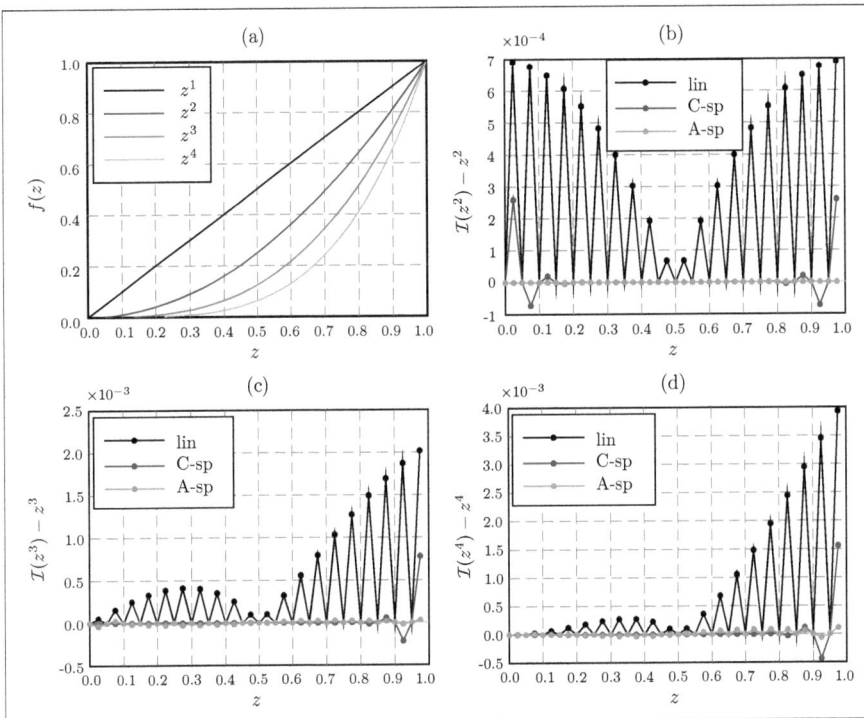

Figure 5.2: Verification of the interpolation procedures: In (a) the imposed distributions are plotted. In (b–d) the resulting interpolation errors for a linear interpolation, and the cubic and AKIMA spline interpolation are plotted. For the linear initial distribution no interpolation errors occur for any of the methods. The linear interpolation is erroneous for any non-linear function. For the quadratic distribution (b) the error of the AKIMA interpolation is zero to machine precision while the error of the C-spline interpolation is very small ($\approx 10^{-9}$) in the middle of the domain but non-zero because of the errors introduced at the domain boundaries and the global character of the spline. For higher order distributions (c,d) all interpolation methods show errors.

5.1.3 Dispersive Behavior of the FIT, LT-FIT and FI-FV scheme

The field pattern of a propagating cylindrical wave, simulated with the FIT, is symmetric with respect to its origin. This should not hold true if the LT-FIT or the FI-FV hybrid scheme is applied instead. Especially if the grid resolution is low, the dispersive behavior along the axes is supposed to show large discrepancies. This is illustrated in the figures 5.3–5.6. There, a cylindrical wave is excited by a line current with a GAUSSIAN time profile. A very low temporal resolution of the exciting current was chosen. Hence, strong dispersion effects are expected to emerge. This test is conducted using the FIT (fig. 5.3) and the FI-FV scheme (fig. 5.4). In both cases, the maximally stable time step is applied. Hence, for the FI-FV scheme no dispersion effects are expected along the z-axis. The result is identical if the FIT is applied to the one-dimensional problem (LT-FIT). However, this does not hold true if any other than the magic time step is applied. In the figures 5.5 and 5.6, half of the magic time step was used with the FI-FV scheme and the LT-FIT method. The improved dispersion properties of the FI-FV scheme along one direction are clearly visible. The application of the split operator procedure in combination with the FIT offers no benefit in this case.

A direct comparison between the dispersion properties of the FIT and the LT-FIT is shown in the figure 5.8. It displays several time instants of a GAUSSIAN wave packet propagating along the z-axis. In both simulations the respective maximally stable time step was chosen. Due to the very low resolution of two grid lines per standard deviation, dispersive effects rapidly emerge during the FIT simulation while the LT-FIT solution shows no dispersion.

5.1.4 Dispersive Behavior of the DGM

The dispersive properties of the DGM are strongly linked to the applied approximation order P (cf. fig. 3.25). In order to verify this, the propagation of a GAUSSIAN wave packet, using identical grid line spacings but different approximation orders is simulated. The results are displayed in the figure 5.7 for orders of $P = 0, 1$ and 2. On the left, the initial distribution of the H_z component is plotted and on the right the final distribution. The wave initially progates in positive x-direction, and it is reflected at the domain boundary. An evaluation of the L^2-error norm of the final distribution for approximation orders up to five yields:

P	0	1	2	3	4	5
L^2	$9.0 \cdot 10^{-1}$	$1.5 \cdot 10^{-1}$	$1.2 \cdot 10^{-3}$	$1.1 \cdot 10^{-4}$	$1.2 \cdot 10^{-5}$	$9.8 \cdot 10^{-6}$

Table 5.1: L^2-error norm for the above example (cf. fig. 5.7) in dependence of the approximation order.

In agreement with the expectations from theorectical investigations the wave shows a phase lag for the approximation using piecewise constant elements and a phase advance for piecewise linear elements (cf. figs. 3.27 and 3.28).

5.1.5 Adaptive FIT Algorithm

The basic functionality of the full adaptive algorithm can be tested using some simple setups. First, it is checked whether the numerical phase velocity increases and reaches the physical value within a refined area. Next, the dispersive behavior of a short GAUSSIAN wave packet, propagating within an adaptively refined region is investigated. For these tests the FIT was employed. The results are illustrated in the figure 5.9. On the left, the normalized numerical phase velocity of a wave front, which is tracked by an adaptively refined region of the computational grid, is shown. Along with the refinement level, the phase velocity increases and asymptotically

5.1. Sanity Check

approaches the physical value (see fig. 3.11). On the right, the electric field along a path in the z-direction for a GAUSSIAN wave packet propagating in this direction is shown. The red curve results from a simulation using no time-adaptive grid refinement and a resolution of three grid lines per standard deviation. Afterwards, the propagating pulse is embedded in a region of local refinement. The resulting curves for refinement levels of $L = 1$, 3 and 5 are shown in the figure. This is an illustrative example where the application of a local time-adaptive grid refinement increases the global accuracy of the solution. The remarkable fact are the savings in memory. If the computational grid is refined globally, the memory consumption increases according to 2^L and yields factors of 2, 8 and 32. The adaptive approach increases the memory demands only by factors of 1.08, 1.32 and 2.13.

5.1.6 Performance

The potential savings in terms of computer memory became obvious in the preceding example. However, it remains to address the computational efforts for performing the grid adaptation. The actual demands depend on the problem under consideration but code profilings have shown that the contribution of the grid refinement procedures to the total CPU time is of the order of a few percent.

5.1.7 Automated hp-Adaptive DG Scheme

The method for performing automated hp-adaptivity, which was presented in section 4.4.4, is applied to the simulation of a GAUSSIAN and trapezoidal wave packet in a one-dimensional domain. Due to the *Convergence Order Theorem* 4.1, it is desirable to choose a high approximation order in regions where the solution is smooth. On the other hand, a small order should be chosen in non-smooth regions. Hence, P should be large in the vicinity of the GAUSSIAN wave packet. For the representation of the trapezoidal packet, constant and linear elements are, in principal, sufficient.

The results of the test are shown in the figure 5.10. The red curve represents the electric field, and the green curve is the magnetic field. The grid lines are indicated as gray dashed lines and the red circles indicate the approximation order used in the respective element. The dynamic range of the approximation order was bound by the lower value of one and the upper value of five. The maximum h-refinement level was prescribed by eight.

In the upper row, the initial distribution and grid are shown. In the middle row, an intermediate solution is plotted and at the bottom, the final solution. The automatic adaptation algorithm succeeds with finding a reasonable hp-grid for all time steps. In the simulation of the GAUSSIAN packet, an h-refinement level of two is not exceeded, while the algorithm takes full advantage of the highest approximation order in the packet region. On the other hand, an approximation order of two is not exceeded during the simulation of the trapezoidal packet. However, the algorithm makes use of h-refinement levels up to six.

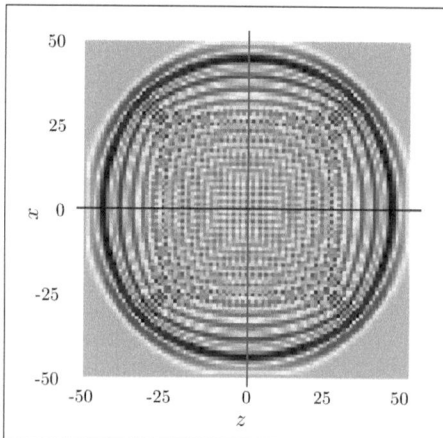

Figure 5.3: Cylindrical wave simulated with FIT: A cylindrical wave is excited by a line current with a GAUSSIAN time profile. The upper plot shows the E_y component in a cut view. A very low temporal resolution of the exciting current was chosen. Hence, strong dispersion effects emerge. The lower plots show the field along the indicated lines in x- and z-direction (a and b). In (c) the difference along the two paths is shown. Evaluating the curl of the magnetic field around the center yields the plot shown in (d) where only the excitation should be visible. The subsequent oscillations are parasitic effects caused by numerical noise.

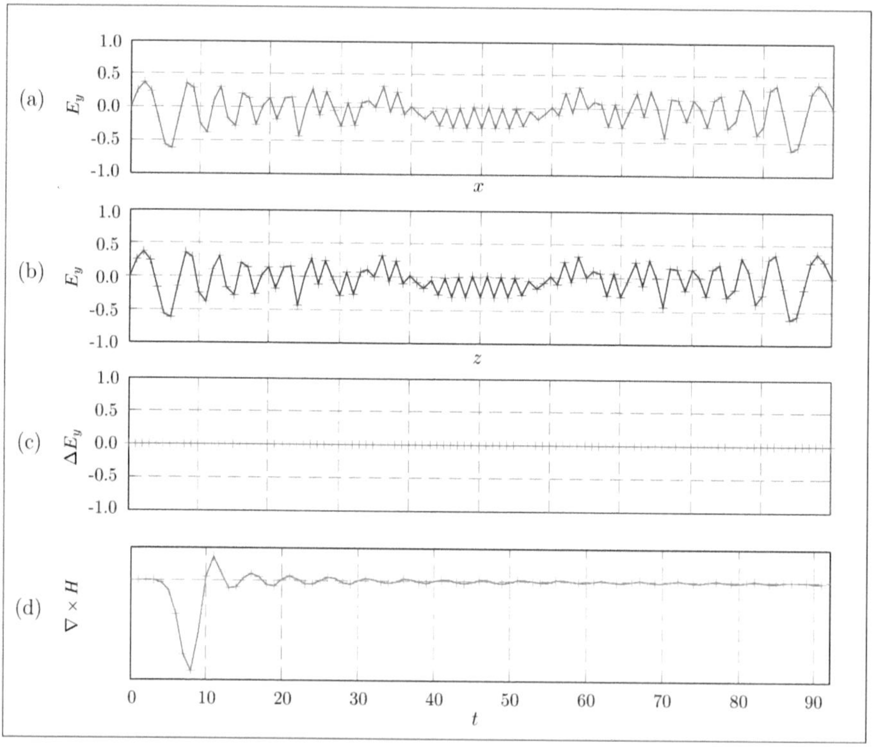

5.1. Sanity Check

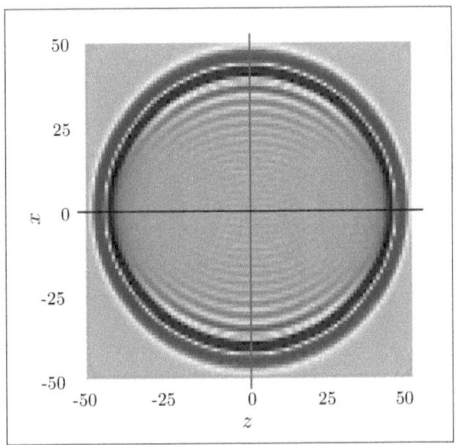

Figure 5.4: Cylindrical wave simulated with FI-FV scheme: The excitation is identical to the preceding figure. The application of the LT splitting scheme limits dispersion effects to the x-axis. The upper plot clearly shows a non-symmetrical propagation. The lower plots show the field along the indicated lines in x- and z-direction (a and b). In (c) the difference along the two paths is shown which is non-zero for this case. In (d) the parasitic oscillations are largely decreased. For the magic time step the FI-FV and the LT-FIT scheme yield identical results.

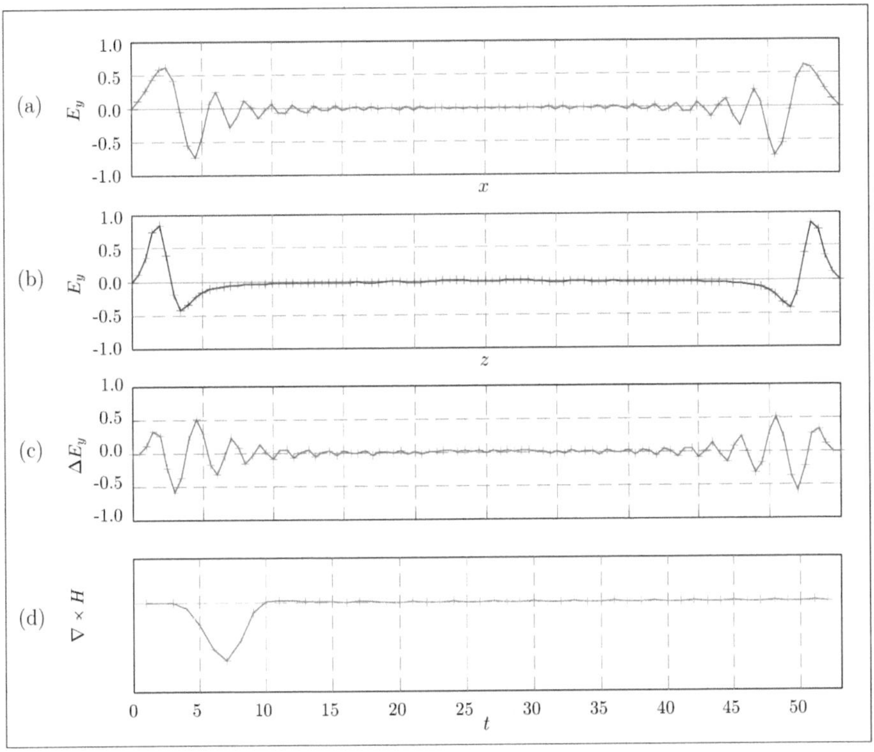

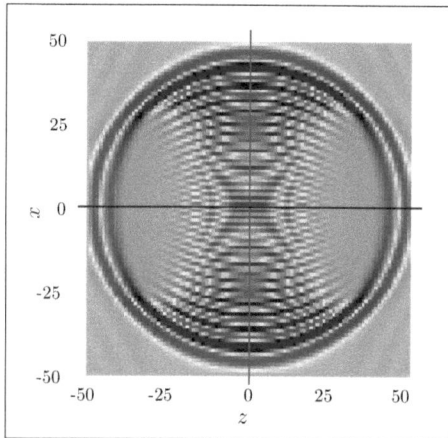

Figure 5.5: Cylindrical wave simulated with FI-FV scheme and reduced time step: The setup is identical to the previous figure except for the applied time step, which is half of the magic time step ($\nu = 0.5$). Dispersion effects occur along both axes, however, much more pronounced along the x-axis. Comparing the maximum amplitude of the plots 5.4b and 5.5b reveals the dissipating character of the FVM for this time step setting. Dissipation is at its maximum for $\nu = 0.5$ (cf. fig. 3.30).

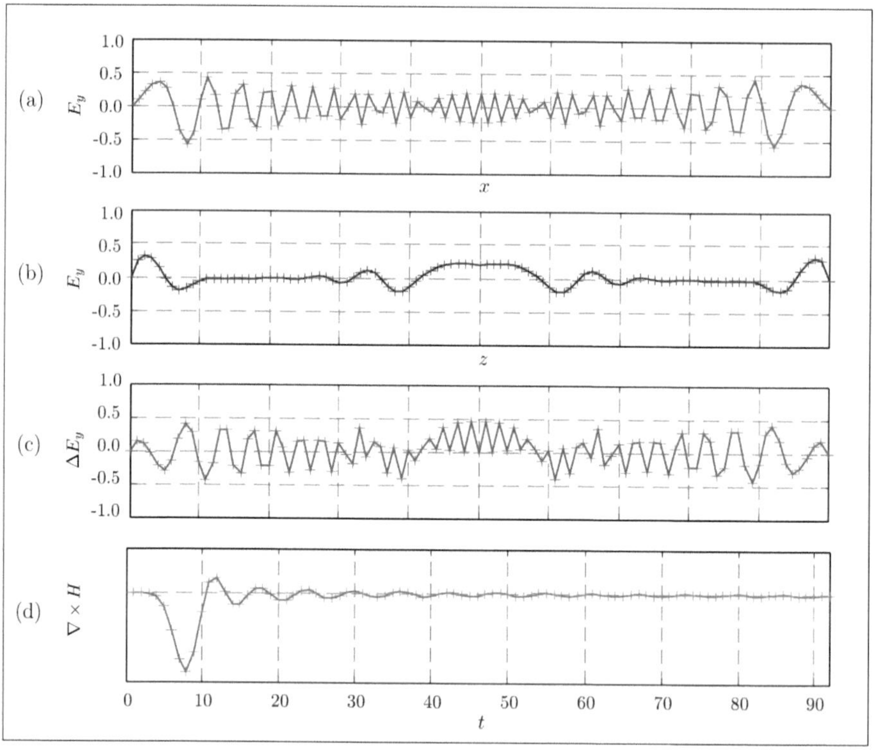

5.1. Sanity Check

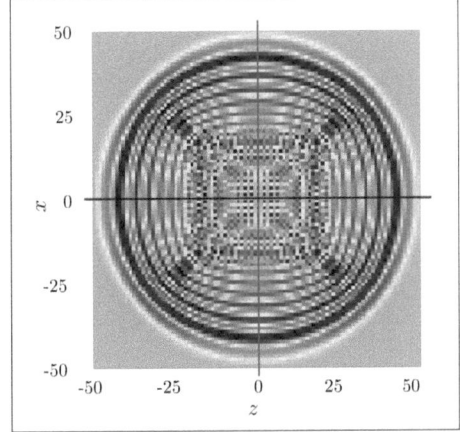

Figure 5.6: Cylindrical wave simulated with LT-FIT and reduced time step: Simulation of the same setup as in the previous figure using the LT-FIT scheme. The application of the LT splitting procedure with the FIT offers no benefit if a time step different from the magic one is chosen ($\nu < 1$).

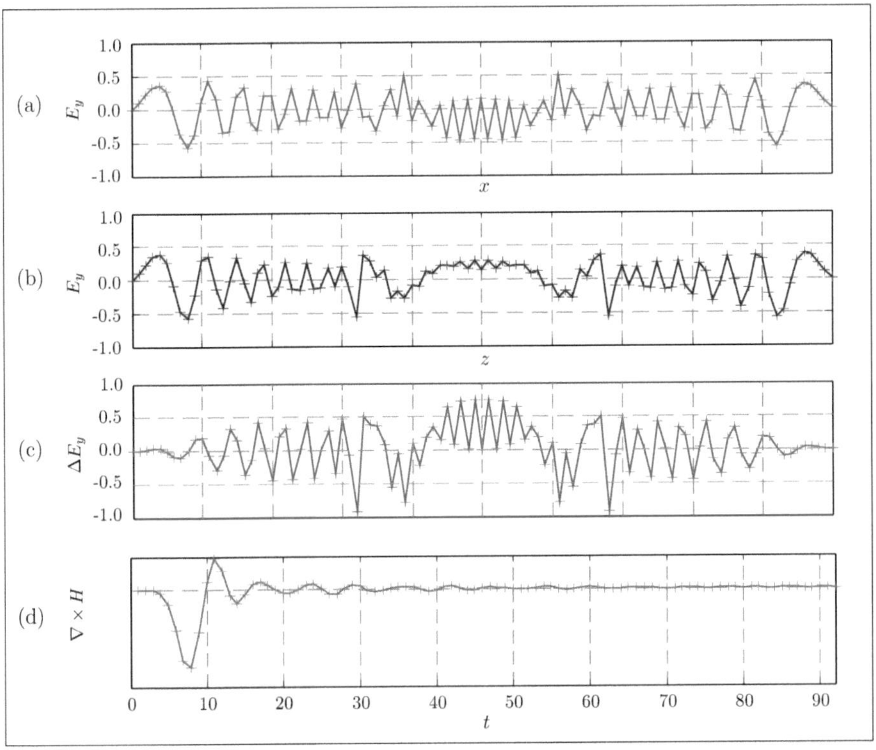

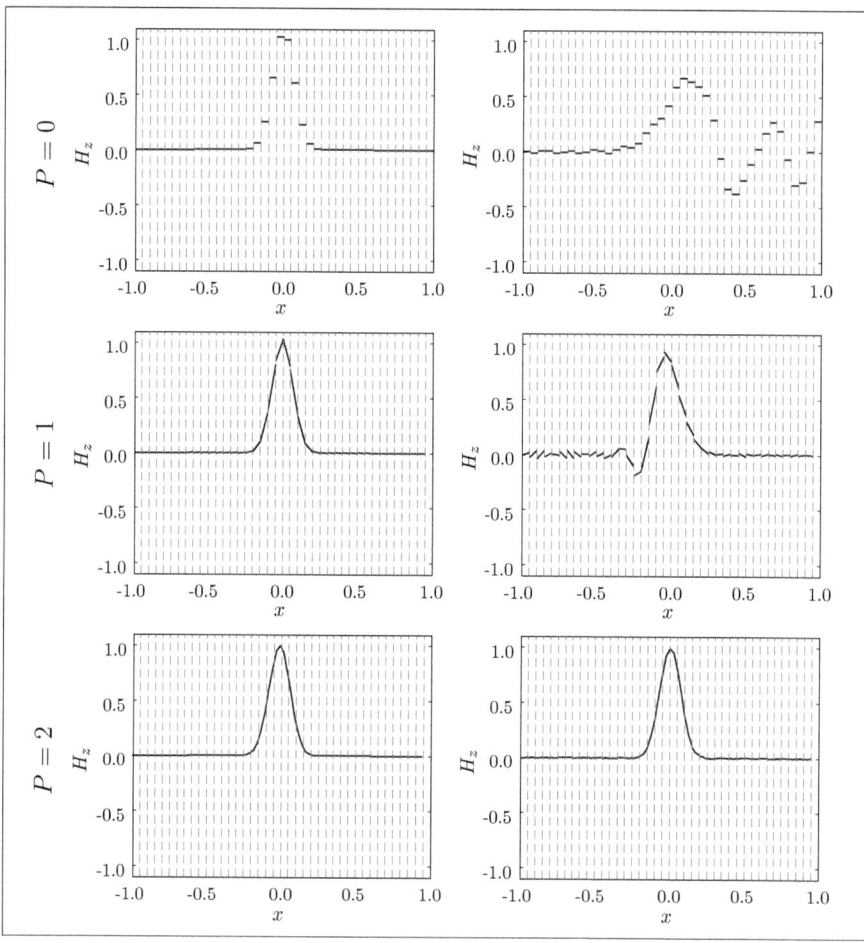

Figure 5.7: Dispersion of the DGM for various approximation orders: The propagation of a GAUSSIAN wave packet along the x-axis is simulated for approximation orders of zero, one and two. On the left, the initial distribution of the H_z component is shown. The wave propagates into the positive x-direction and is reflected at the border of the computational domain. The final distribution is shown in the right column. For piecewise constant elements, the wave shows a phase lag and for piecewise linear elements a phase advance (cf. figs. 3.27 and 3.28).

5.1. Sanity Check

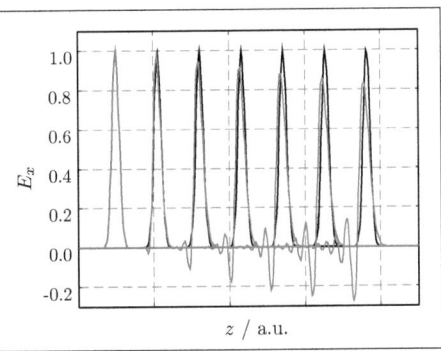

Figure 5.8: Dispersion of FIT and LT-FIT: Propagation of a GAUSSIAN wave packet. The plot shows snapshots of the x-component of the electric field along a z-directed path. The black curve results from the LT-FIT, for the gray curve the standard FIT was used. For both simulations the respective maximum time step was applied. The resolution of the pulse is two grid lines per standard deviation σ.

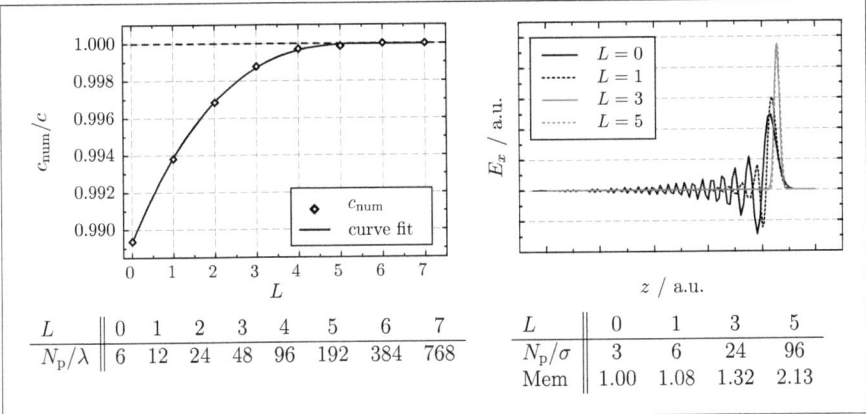

L	0	1	2	3	4	5	6	7
N_p/λ	6	12	24	48	96	192	384	768

L	0	1	3	5
N_p/σ	3	6	24	96
Mem	1.00	1.08	1.32	2.13

Figure 5.9: Numerical phase velocity (left) and noise level (right) vs. refinement level (FIT): On the left it is illustrated how the numerical phase velocity approaches the physical value if grid refinement is applied. The test was conducted by refining the grid in the vicinity of a wave front by the refinement factors $L = 0..7$. In the base grid, the wavelength λ was resolved by six grid lines (N_p). The resolution is doubled with every increase of the refinement level.
On the right, the effect of the resolution dependent phase velocity is illustrated with the propagation of a GAUSSIAN wave packet. For low resolutions strong dispersive effects are observed (red) which nearly vanish for high resolutions (blue). In this example, the packet was embedded in a region of adaptive grid refinement using the levels 0,1,3 and 5. The resulting numbers of grid lines per standard deviation σ of the packet are listed in the table. The remarkable result is the relative increase of required computer memory (Mem) with respect to the non-refined case $L = 0$. If the grid resolution is consecutively doubled within the complete computational domain, the required amount of memory scales with 2^L. This yields factors of 2, 8 and 32 for the given refinement levels. For the adaptive approach, on the contrary, these factors are 1.08, 1.32 and 2.13.

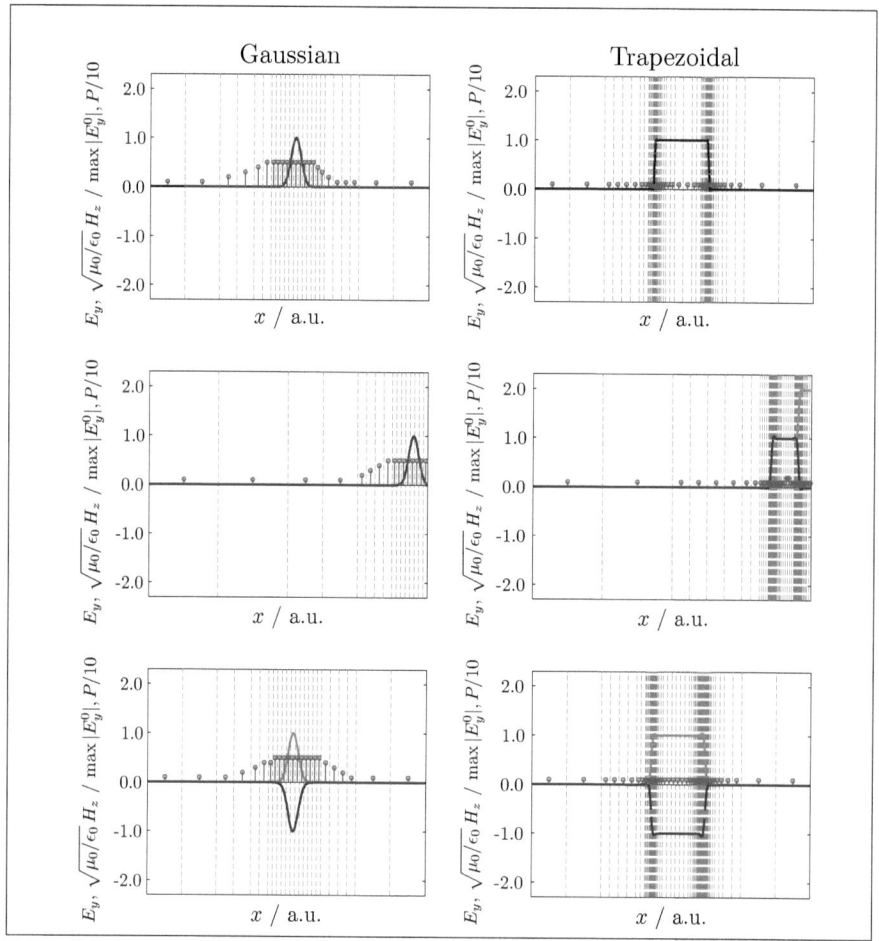

Figure 5.10: Simulation of a GAUSSIAN and trapezoidal wave packet using the automated hp-adaptive algorithm: The electric field is plotted in black and the magnetic field in gray. The grid lines are indicated as gray dashed lines and the circles indicate the approximation order used in the respective element. The dynamic range of the approximation order was bound by the lower value of one and the upper value of five. The maximum h-refinement level was prescribed by eight. The initial, an intermediate and the final solutions are shown from top to bottom. In the simulation of the GAUSSIAN packet, the h-refinement does not exceed a level of two, while the algorithm takes full advantage of the highest approximation order in the packet region. For the trapezoidal packet, an approximation order of two is not exceeded. However, the algorithm makes use of h-refinement levels up to six.

5.2 Comparison and Benchmarking of the Numerical Methods

In this section, a comparison of the numerical methods is carried out. They are applied to the simulation of a non-relativistic particle bunch in a circular metallic tube. The settings of the computational model are specified in the table below. In [92] the analytical solution of this problem for a semi-infinite tube is given.

Tube length	Tube radius	RMS bunch radius	RMS bunch length	Particle velocity
120 mm	40 mm	5 mm	3 mm	$0.9\,c_0$

Table 5.2: Settings of the benchmark example

In the simulations the axis of the cylinder is aligned with the z-coordinate. In the plots (a) and (c) in figure 5.11, the analytical solution of the longitudinal electric field along the axis is depicted by a black curve. The solutions obtained with the FIT, the LT-FIT, the FI-FV scheme using the GODUNOV (FI-FV1) and the FROMM (FV-FI2) slope, and the DGM using piecewise linear elements are compared to the analytical solution. For all results shown in (a), the same equidistant grid and the respective maximum time step was applied. In (b), the error is plotted. For this case, the solutions of the LT-FIT and the FI-FV1/FI-FV2 are identical. The results obtained using time-adaptive grid refinement ($L = 3$) in the bunch region are shown in (c) and the errors in (d). There, the FI-FV1 scheme shows the largest error. The magic time step can be applied only in the maximally refined region. In all other regions, its pronounced dissipative character leads to large amplitude errors (see fig. 3.30).

A series of simulations of the above problem using various grid and grid refinement settings was carried out. The results of this study are presented in the tables 5.3–5.6. Besides the grid settings, they list the number of DoF applied and the required time in seconds. The relative L^2-error and the *total variation* (TV) of the longitudinal electric field along the cylinder axis are given as measures for the quality of the numerical solutions. The relative L^2-error is computed as:

$$\varepsilon^{\text{rel}} = \frac{\|\mathbf{E}_z - \mathbf{e}_z\|_2}{\|\mathbf{E}_z\|_2} \quad \text{with} \quad \mathbf{E}_z = \Big(E_z(z(1)), .. E_z(z(N_z))\Big)^{\text{T}} \tag{5.1}$$

where E_z is the longitudinal component of the analytical solution for the electric field, and $z(i_z)$ denotes the respective grid point positions. The vector $\mathbf{e}_z$ contains the numerical solution along the axis obtained with the respective method. For the FIT and FI-FV scheme, it holds one sampled value per edge, for the DGM it contains the approximate solutions evaluated in the cell centers.

The total variation is a measure for the smoothness of a function [63]. It is defined as:

$$TV(E(z)) := \int |E'(z)|\,\mathrm{d}z \tag{5.2}$$

In [77] it is shown that the TV of a function equals the sum of weighted local minima and maxima. It is, therefore, a direct measure for oscillations and their amplitudes. The TV of discrete solutions is computed as:

$$TV(\mathbf{e}_z) = \sum_{i_z=1}^{N_z} |e_z(i_z+1) - e_z(i_z)| \tag{5.3}$$

The tables contain the results obtained with the FIT (5.3), the LT-FIT (5.4), the FI-FVM (5.5), and the DGM (5.6). Results obtained with the standard algorithm on static grids are listed first. Then, the results obtained with time-adaptive grid refinement are given. In that case, N_z is the number of grid points in the non-refined base grid along the z-axis. The refinement level is chosen such that the grid resolution in the bunch area is approximately the same in all cases.

The LT-FIT and FI-FV scheme yield identical results when applied on equidistant grids using

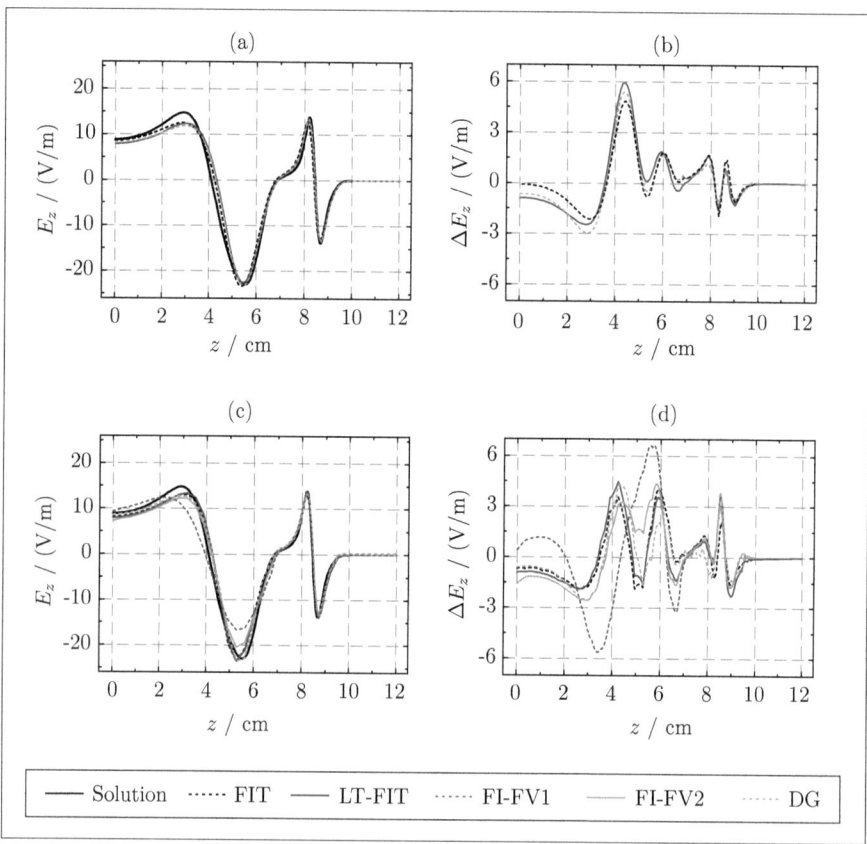

Figure 5.11: Longitudinal component of the electric field, excited by a GAUSSIAN bunch traveling in a tube: The black curve in (a) and (c) indicates the analytical solution for the setup described in the table 5.2. All other curves shows results obtained by the numerical simulation of the problem. In the top row, a fixed computational grid was applied. The results are shown in (a), their error in (b). In the bottom row, time-adaptive grid refinement ($L = 3$) was applied. In (c) the results for this case are shown and in (d) the errors.

5.2. Comparison and Benchmarking of the Numerical Methods

the magic time step. Therefore, only results obtained with the time-adaptive method are presented for the FI-FV. In the upper half of the respective table 5.5, GODUNOV slopes were applied. The results given in the lower half were obtained using FROMM's slope definition. For the DGM, results obtained with h-, p- and hp-refinement are shown in the table 5.6.

The FIT is considered to be the most common method for performing PIC simulations. All TV values are, hence, normalized to the value obtained with the FIT on the finest considered grid. For comparability, the settings were chosen such that the simulations finish within less than one hour (with one exception in table 5.6).

A comparison of the results obtained on static grids shows that:

- the relative errors of the FIT and LT-FIT are similar, however, the TV of the LT-FIT solution is significantly lower, which indicates reduced numerical noise. While the numbers of DoF for both methods are identical, the LT-FIT simulations take longer because of the increased costs for the time integration.

- the required time for computing a DGM solution using piecewise constant basis functions is about half the time consumed for solving it with the FIT. This stems from the maximally applicable time step, which is twice as large for the DGM than for the FIT. However, the relative errors and the TV of the DGM solutions for $P = 0$ are much larger.

- the error as well as the TV of the DGM solutions largely decrease, if the approximation order is increased. This agrees with the theoretical results presented in the sections 3.4 and 3.6. However, also the number of DoF and the simulation time are strongly linked to the approximation order. The DGM using piecewise constant elements on a grid of $135 \times 135 \times 210$ nodes, and piecewise quadratic elements on a grid $45 \times 45 \times 35$, employ the same number of DoF, which is approximately $22.96 \cdot 10^6$. Nevertheless, the error for the latter case is about a factor of five smaller, and the TV is about a factor of two smaller. Hence, the higher convergence order, connected with high order basis functions, outweighs the associated higher computational costs and leads to more accurate results.

- for a number of $22.96 \cdot 10^6$ DoF the error of the DGM solution using piecewise quadratic elements is the smallest when compared to the respective errors of the FIT and LT-FIT solutions employing the same number of DoF. The LT-FIT achieves the least TV.

Comparing the results obtained on time-adaptive grids shows that:

- the relative errors are comparable for comparable minimum grid step sizes. Hence, the relative errors of the solutions are mainly determined by the resolution in the bunch region. For each of the methods, the relative error is approximately the same for all refinement settings ($L = 1..4$).

FIT

L	N_x	N_z	Δx / mm	$\Delta z_{\min}$ / mm	DoF / 1e6	Time / sec	$\mathcal{E}^{\text{rel}}$	TV
				static grid				
0	135	210	0.59	0.57	22.96	2066	0.138	1.00
0	91	140	0.88	0.86	6.96	386	0.165	1.82
0	45	70	1.78	1.71	0.85	31	0.251	3.27
0	23	35	3.48	3.43	0.11	1	0.426	7.05
				time-adaptively refined grid				
1	135	105	0.59	0.57	13.00	1225	0.145	1.42
2	135	52	0.59	0.58	9.00	940	0.147	1.93
3	135	27	0.59	0.56	8.50	925	0.147	1.99
4	135	13	0.59	0.58	11.50	1025	0.130	1.55

Table 5.3: Results of FIT simulations: The upper half shows results for equidistant static grids, in the lower half the results obtained on time-adaptive grids are displayed.

LT-FIT

L	N_x	N_z	Δx / mm	$\Delta z_{\min}$ / mm	DoF / 1e6	Time / sec	$\mathcal{E}^{\text{rel}}$	TV
				static grid				
0	135	210	0.59	0.57	22.96	3543	0.138	0.63
0	91	140	0.88	0.86	6.96	626	0.159	1.31
0	45	70	1.78	1.71	0.85	31	0.259	2.99
0	23	35	3.48	3.43	0.11	1	0.336	5.93
				time-adaptively refined grid				
1	135	105	0.59	0.57	13.00	2008	0.137	1.39
2	135	52	0.59	0.58	9.00	1443	0.141	1.65
3	135	27	0.59	0.56	8.50	1447	0.140	1.80
4	135	13	0.59	0.58	11.50	1791	0.119	1.40

Table 5.4: Results of LT-FIT simulations: The upper half shows results for equidistant static grids, in the lower half the results obtained on time-adaptive grids are displayed.

FI-FVM

L	N_x	N_z	Δx / mm	$\Delta z_{\min}$ / mm	DoF / 1e6	Time / sec	$\mathcal{E}^{\text{rel}}$	TV
				GODUNOV				
1	135	105	0.59	0.57	13.00	1421	0.228	0.90
2	135	52	0.59	0.58	9.00	1070	0.256	0.84
3	135	27	0.59	0.56	8.50	1033	0.283	1.04
4	135	13	0.59	0.58	11.50	1217	0.276	1.06
				FROMM				
1	135	105	0.59	0.57	13.00	1423	0.123	0.95
2	135	52	0.59	0.58	9.00	1092	0.132	0.90
3	135	27	0.59	0.56	8.50	1048	0.153	1.02
4	135	13	0.59	0.58	11.50	1233	0.136	1.06

Table 5.5: Results of FI-FV simulations: The upper half shows results for time-adaptive grids using GODUNOV slopes. The results shown in the lower half were obtained using FROMM's slope.

5.2. Comparison and Benchmarking of the Numerical Methods

DGM

L	P	P_{fine}	N_x	N_z	Δx	$\Delta z_{\min}$	DoF / 1e6	Time / sec	$\mathcal{E}^{\text{rel}}$	TV
\multicolumn{11}{c}{static grid}									
0	0	0	135	210	0.59	0.57	22.96	1194	0.413	2.53
0	0	0	91	140	0.59	0.85	6.96	234	0.663	4.38
0	0	0	45	70	0.59	1.71	0.85	16	1.636	9.98
0	0	0	23	35	0.59	3.43	0.11	1	3.119	10.57
0	1	1	91	140	0.59	0.85	55.65	4102	0.136	1.10
0	1	1	45	70	0.59	1.71	6.80	311	0.163	1.62
0	1	1	23	35	0.59	3.43	0.88	19	0.381	2.00
0	2	2	45	70	0.59	1.71	22.96	2018	0.083	1.15
0	2	2	23	35	0.59	3.43	2.99	200	0.142	1.63
\multicolumn{11}{c}{time-adaptive h-refinement}									
1	1	1	45	35	1.78	1.71	4.00	216	0.171	1.44
2	1	1	45	18	1.78	1.66	3.50	180	0.183	1.55
1	2	2	45	35	1.78	1.71	14.00	1171	0.084	1.21
2	2	2	45	18	1.78	1.66	11.50	1672	0.086	1.36
\multicolumn{11}{c}{time-adaptive p-refinement}									
0	1	2	45	70	1.78	1.71	8.50	455	0.128	1.39
0	1	3	45	70	1.78	1.71	13.00	987	0.061	0.88
\multicolumn{11}{c}{time-adaptive hp-refinement}									
1	2	1	45	35	1.78	1.71	13.00	1171	0.113	1.44
2	3	1	45	18	1.78	1.66	9.00	826	0.101	1.35

Table 5.6: Results of DGM simulations: On top, results obtained with equidistant static grids and approximation orders from zero to two are shown. Below, time-adaptive grid refinement using h-, p- and hp-refinement was applied.

- the relative errors obtained with the FI-FV scheme applying GODUNOV slopes are significantly larger than the errors obtained with any other method. This is due to its strongly dissipative character for any but the magic time step, which results in large amplitude errors.

- the FI-FV scheme yields the smallest total variations. This confirms the improved dispersion properties, which were expected from its theoretical investigations presented in the sections 3.3 and 3.6.

- the FI-FV scheme consumes less computational time than the LT-FIT method and only slightly more than the FIT. The extra time consumed for evaluating the FROMM slopes is negligible.

- the smallest error is obtained for the DGM applying pure p-refinement. For the given velocity of the particles, the LORENTZ contraction of the excited self-fields plays a minor role and the field distribution is smooth. As discussed in section 4.4.4, p-refinement is the most efficient refinement technique in this case.

5.3 Self-Consistent Simulation of the PITZ RF Gun

The PITZ RF Gun

The injector for FLASH and the XFEL is developed in the framework of the PITZ project (Photo Injector Test Facility at DESY Zeuthen) [53]. Its purpose is the development and testing of a high quality, pulsed electron source. Besides the high quality concerning beam emittance, the injected electron bunches have to be highly charged in order to achieve the phase space density required for the self-amplified spontaneous-emission (SASE) [70] to occur in the downstream Free-Electron Laser (FEL). The extraction of a bunch charge of 1 nC from an emission area of approximately 3 mm^2 and a duration of about 20 ps leads to a space charge dominated beam with local current densities above 1 kA/cm^2. This exceeds the limit which can be provided by thermal cathodes. Therefore, laser-driven photo cathodes are used.

The layout of the injector has been shown in the figures 1.1 and 1.2. The emitted bunch is accelerated in a 1.5-cell L-band cavity providing for an accelerating gradient of 42 MV/m at an operation frequency of 1.3 GHz. In a later stage of the project, it is planned to raise the gradient to 60 MV/m. The space charge induced emittance growth is partially caused by a correlation of the transverse phase space distribution and the longitudinal position within the bunch. This can be corrected using a focusing scheme proposed in [29]. The gun layout is designed such that a minimum of the transverse emittance is expected at a distance of approximately 1.6 m from the cathode. At this position another RF cavity will be installed in order to accelerate the electrons to relativistic energies. This is expected to conserve the minimum of the transverse emittance. The main design parameters are listed in the table 5.7.

Parameter	Design value
Bunch charge	1 nC
Transverse laser profile	Hat-profile
Laser spot radius	1 mm
Longitudinal laser profile	Flat-top
Laser pulse duration	22 ps
Rise/Fall time	2 ps
Accelerating gradient	42 MV/m
Transverse emittance	1 mm mrad

Table 5.7: PITZ design parameters

Many numerical studies of the injector using PIC codes have been performed over the last years. Results have been published, e.g., in [30, 106, 108, 116, 149]. However, these simulations either assumed a rotationally symmetric geometry and made use of a so-called 2.5-dimensional approach as implemented in the MAFIA TS2 module [40], or distances of only a few cm of the full three-dimensional model were simulated. In 2006, results of parallelized 3D simulations up to 1 m downstream of the cathode have been presented in [57, 143]. However, the exact position and value of the transverse emittance minimum are required for the optimal placement of the following accelerating cavity.

In the context of this work, the gun was simulated up to a distance of 2 m downstream of the cathode. First, parameter settings such as the grid resolution and the number of macro

5.3. Self-Consistent Simulation of the PITZ RF Gun

particles have been determined. Then, a design study was performed, which addressed the effects of individual injector elements on the beam quality.

Parameter Settings

Grid step sizes: Previous investigations have shown that the length of the simulated bunches critically depends on the longitudinal grid step size [107]. In [149] it was stated that grid resolutions of 20 μm are required in order to obtain accurate results. In [108], a parameter study addressing the relation of the grid step size and the computed bunch length was presented. The results are shown in the figure 5.12a. The initial 2.5 cm of the gun were simulated using an equidistant grid with a step size of 2.5 μm to 80 μm. The computed RMS bunch length, σ_z, at $z = 2.5$ cm varies from 2.24 mm to 2.35 mm. The deviation in σ_z for a step size of 2.5 μm and 5 μm is around 1 μm. The computed bunch lengths for a step size of 2.5 μm and 10 μm differ by approximately 10 μm. Hence, a longitudinal grid step of 10 μm is considered to be a reasonable choice.

After their emission, the electrons have a very low energy of 5 eV and space charge forces have the strongest influence. Since the particles quickly gain energy, it is sufficient to apply the smallest step size only in the vicinity of the cathode. Hence, an additional static grid refinement in this area was applied. In the figure 5.12b the results are shown. There, a step size of 80 μm for the base grid, and refinement levels from one to four were chosen. The static refinement was applied within 1 cm distance from the cathode. The results are, except for minor discrepancies, identical to those obtained with an equidistant grid. For the simulation of the full structure, the longitudinal grid step combination 10 μm/80 μm was applied.

The requirements on the transverse grid step size are less demanding. The width of the simulated bunches using a step size of 2.5 μm and 80 μm differ by less than 5 μm. Deviations in the μm-range correspond to errors of some per mille.

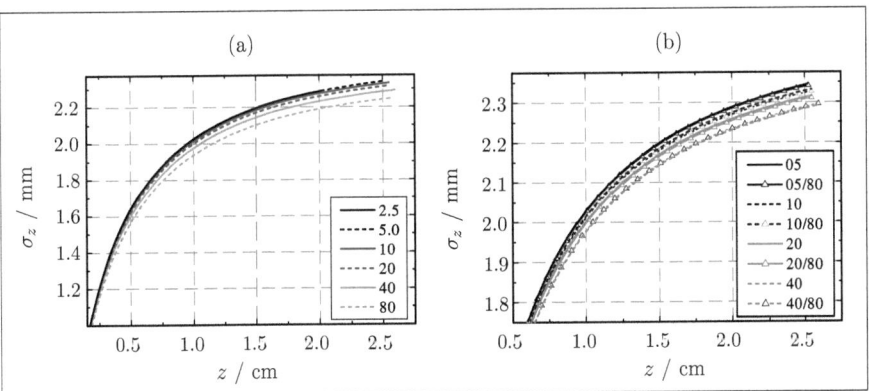

Figure 5.12: Computed RMS bunch length vs. longitudinal grid step size: In (a) equidistant grids with different grid step sizes were used. In (b) an equidistant base grid with a step size of 80 μm was statically refined within the first cm from the cathode. The step sizes given in the legend are in μm.

Number of emitters per cell: The length and width of the simulated bunches also depend on the number of macro particles in the bunch. In the developed code, the number of discrete particle emitters per cell is specified instead of the total number of particles to be emitted. In order to obtain accurate and low-noise convective currents in the computational grid, every cell has to be occupied by several macro particles. Specifying the number of discrete particle emitters within every cell in the cathode area makes it easier to ensure this. For the grid settings given above, the resulting RMS bunch width and length for numbers of one to 32 emitters per cell are shown in the figure 5.13. For the simulation of the full structure eight emitters have been selected.

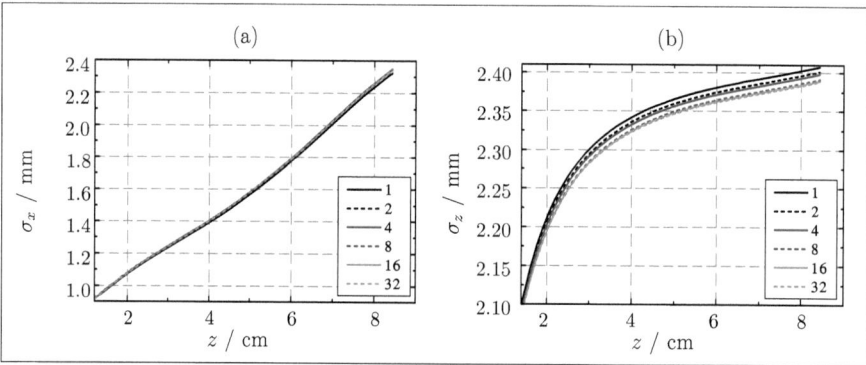

Figure 5.13: Computed RMS bunch width and length vs. number of emitters per cell: In (a) the RMS bunch width and in (b) the RMS bunch length for a number of 1 to 32 emitters per cell is shown.

Design Study

The PITZ RF gun was simulated applying the presented techniques. However, the results given in the following were obtained with the FI-FV scheme employing FROMM's slopes. In the figure 5.14, the evolution of the RMS bunch width (a), the RMS bunch length (b), the average particle energy (c), and the emittance ε^m, calculated from mechanical momenta, (d) are plotted. In the following the superscript for the emittance is dropped.

The results of the design study are shown in figure 5.15. The purpose of this study is to specify the effects of the diagnostics doublecross, the laser mirror, and the shutter valve on the beam quality in terms of emittance growth. The respective models are shown in the figures 5.15a–d. For each model, one element is added. The colors of the result curves in 5.15e–f correspond to the outline color of the respective model.

For the model (a), the horizontal and vertical emittances, ε_x and ε_y, are equal. This is expected since the geometry is rotationally symmetric except for the RF coupler. In (b), the large opening on the bottom side of the doublecross introduces an asymmetry of the horizontal and vertical geometry. While ε_x is almost unchanged, the value of ε_y increases by approximately 0.04 mm mrad. The laser mirror, added in the horizontal plane, introduces another asymmetry (c). Its influence on the transverse emittances is visible in the result curves but the emittance growth is rather small. Finally, the inclusion of the shutter valve in (d) results in significant changes of the transverse emittances. While the horizontal emittance actually decreases by approximately 0.1 mm mrad, the vertical emittance increases about 0.18 mm mrad in comparison to case (a). Hence, the shutter valve causes a distinct asymmetry in the transverse bunch distribution. This asymmetry has been addressed in [105], where RF shields in the valve area where investigated using the PBCI wake field code [56].

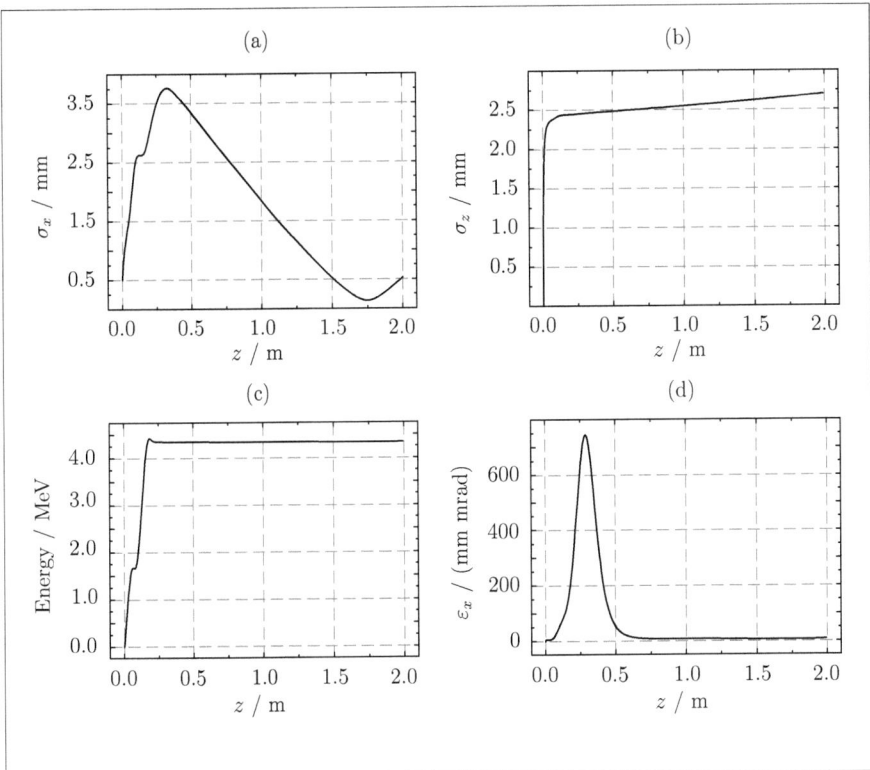

Figure 5.14: Evolution of bunch parameters in the PITZ RF gun: The plots show the evolution of the RMS bunch width (a), RMS bunch length (b), average particle energy (c), and horizontal emittance (d) along the longitudinal coordinate.

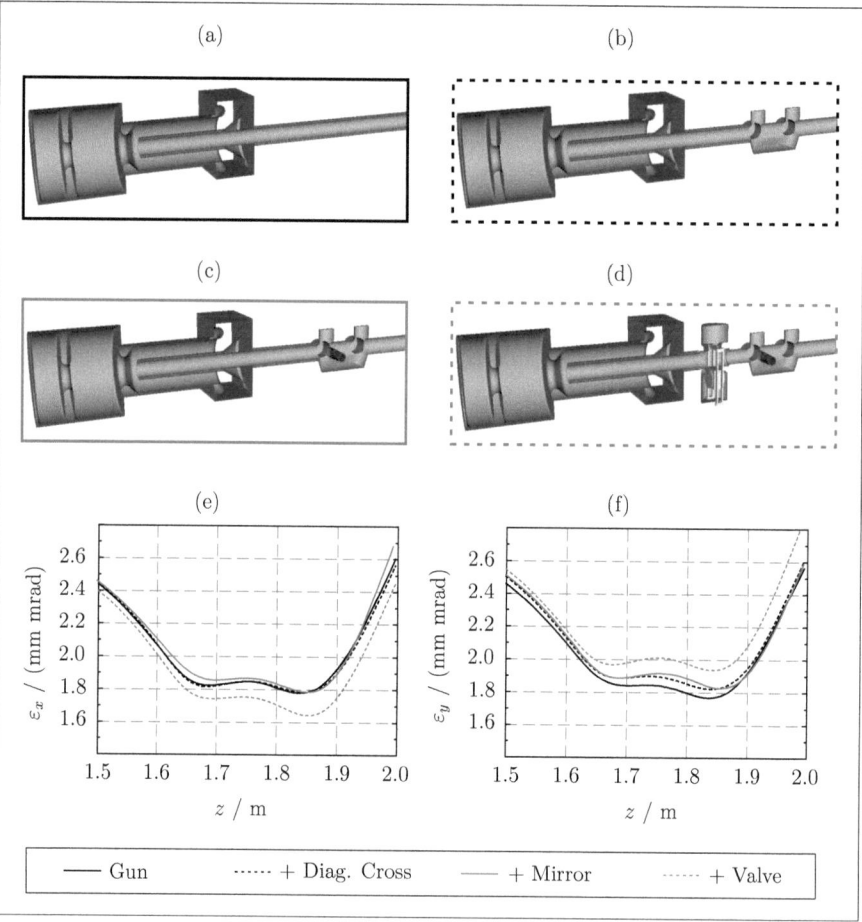

Figure 5.15: Models and results of the PITZ gun design study: In (a) to (d) the CAD models of the simulated structures are shown. The horizontal and vertical emittances around the position of their minimum are plotted in (e) and (f) respectively. The grayscale and line art of the result curves correspond to the outline of the models above. While ε_x and ε_y are similar for the case (a), they show a distinct asymmetry for the model (d).

6. SUMMARY AND OUTLOOK

In this thesis, time-adaptive grid refinement techniques allowing for efficient and accurate self-consistent beam dynamics simulations have been established.

After stating the fundamental equations of MAXWELL and NEWTON-LORENTZ in continuum (chapter 2), techniques for the spatial discretization of the electromagnetic quantities have been presented in chapter 3. These methods are the Finite Integration Technique, a new hybrid Finite Integration–Finite Volume scheme, and the Discontinuous GALERKIN Method. The semidiscrete equations have been investigated for their numerical dispersion and, if existent, numerical dissipation properties. Although the formal definition of consistency was retained for a later discussion, it has already been concluded that all methods are consistent in the sense that their dispersion and dissipation errors tend to zero if the grid step size is continuously decreased. This behavior is illustrated in the dispersion graphs 3.11 (FIT), 3.15 (FIT–FVM), and 3.18, 3.19 (DGM). From these graphs, it is evident that the key to reduced dispersion errors is a high resolution of the computational grid. However, for multi-scale problems, such as the presented FLASH injector, the global application of a very fine grid is not feasible because of computer memory and simulation time limitations. Since the very high-frequency fields are excited by the bunch and concentrated in its vicinity, it is sufficient to restrict the high resolution to this region. Hence, chapter 4 was concerned with techniques for performing time-adaptive grid refinement. In the following the development is reviewed and the main achievements are highlighted.

In the framework of the FIT, time-adaptive grid refinement is a new development. For the FVM and the DGM, such techniques have been published (e.g. in [13, 28, 45]) but their combination with the self-consistent simulation of charged particle dynamics has not been reported on.

Within the FIT and the FI-FV scheme, the discrete field quantities of added or modified grid cells are determined by the interpolation of neighboring quantities. For this task, linear as well as third order spline interpolations have been investigated. While splines offer a higher accuracy, their setup usually requires the solution of a system of equations. For long-term simulations of transient phenomena, this leads to a considerable computational overhead. In addition, RUNGE's phenomenon of overshooting can introduce undesired oscillations. Both issues have been tackled with the employment of AKIMA sub-spline functions. Their setup requires only local information and the reduced continuous differentiability of sub-splines largely mitigates RUNGE's phenomenon. The additional incorporation of slope limiting techniques with the AKIMA splines resulted in a computationally inexpensive method for performing third order spline interpolations, which are entirely void of overshoots.

The options for adapting the local resolution within the DGM are twofold. Besides an actual manipulation of the computational grid (h-refinement), the space of basis functions can be enriched or reduced (p-refinement). For both types, techniques based on the projection of the current approximate solution to the space of basis functions associated with the h- or p-adapted element have been presented. It has been shown that the projection to h-refined and h-coarsened elements can be expressed by means of matrix operators. These operators can be analytically evaluated and stored. This allows for highly efficient h-adaptations during the actual time-domain simulation. Performing p-adaptations for the employed set of orthogonal basis functions has been shown to be mathematically trivial. It is achieved by enriching or reducing the basis of the respective cell. In case of an enrichment, the new coefficients are initialized by zero, and in case of a reduction the highest order coefficients are removed. All other coefficients are not affected. The optimality of the projection based adaptation techniques in the sense of the *Optimality Theorem* 3.2 was concluded.

Each of the presented adaptation techniques has been proven to preserve the stability of the non-adaptive basic algorithm. For the FIT and the FI-FV scheme a continuous energy density, based on the discrete electromagnetic quantities, has been introduced. It was shown that the total energy in the computational domain, obtained by integrating this density, is a conserved quantity during grid adaptation for the applied interpolation techniques. The stability of the projection based adaptation techniques for the DGM has also been proven. It was shown that h-refinement and p-enrichment preserve a consistently defined energy of the discrete electromagnetic field solution exactly. Performing h-coarsening or p-reduction, on the contrary, does, in general, lead to a loss of energy.

The combination of h- and p-refinement yields an hp-adaptive scheme. These schemes are highly attractive because exponential convergence rates have been established in their theoretical analysis [45]. A scheme for performing error controlled, automated hp-adaptive simulations has been presented. The decision upon h- or p-adaptation is based on estimating the local smoothness of the approximate solution. The assignment of a local resolution to every cell follows a heuristic approach. This approach was published in [99] and complemented by some constraints in order to render it more robust. The variable global number of DoF, but especially the adaptation of the local number of DoF during p-adaptations, make the implementation of an efficient hp-adaptive scheme a non-trivial task. The frequent reallocation of memory requires an efficient memory management strategy. Such a strategy was implemented, and it is outlined in the appendix C.

For the time-domain simulation of multi-scale problems, such as the FLASH injector, very large numbers of time steps have to be performed. In this case, the accumulation of dispersion errors demands for prohibitively fine grid resolutions that can render the problem unmanageable, even if time-adaptive grid refinement is applied. Split operator techniques allow for the elimination of dispersion errors along a coordinate axis. Despite the fact that the very high-frequency fields, excited by the particle bunch, still require a high resolution in the vicinity of the bunch, the lack of dispersion errors within the maximally refined area relaxes the demands on the grid resolution.

Besides the main goal of developing techniques for performing time-adaptive grid refinement in the framework of self-consistent beam dynamics simulations, some further theoretical achievements have been accomplished. They are:

- The development of a hybrid FI-FV scheme based on the longitudinal-transversal (LT) split operator technique. The scheme is free of dispersion errors along one coordinate if the magic time step is applied. For any other stable time step, the dispersive behavior is superior to the exclusive application of the FIT with the LT splitting technique. Analytical formulae for describing the semidiscrete dispersive and dissipative properties of the GODUNOV, LAX-WENDROFF and FROMM FV method have been derived. Asymptotic convergence orders have been established. In addition, an accuracy analysis of the fully discretized equations for various CFL numbers has been presented.

- It was highlighted that the DGM applied on Cartesian grids with a tensor product basis yields a very efficient, high order numerical method. For this setting, the mass matrices reduce to a strictly diagonal form. Their required inversion, which is the main issue of other FEM formulations in time-domain applications, is, thus, a trivial task. Moreover, it was shown that the elements of the derivative matrices can be evaluated analytically and stored at negligible memory costs. For the application of central fluxes the method conserves the discrete energy, and it is void of parasitic charges. The latter property is of vital importance for the coupled simulation of electromagnetic fields and charged particle dynamics. An analysis of the dispersive properties of the semidiscrete and the fully discretized DG equations was carried out. Stability limits for approximation orders up to five have been presented.

- The immediate connection between the eigenvalue distribution of the system matrix, the conservation of the discrete energy, and applicable time integration schemes was highlighted. In this context, it was shown that the LT operator splitting using GODUNOV's scheme is always unstable for three-dimensional problems while the STRANG scheme provides a stable technique if the magic time step is applied.

- A reformulation of the PIC method of VILLASENOR and BUNEMAN [134] has been presented which is charge preserving for arbitrary particle motions in three-dimensional space.

From a practical point of view the main achievements of this thesis are:

- The development of a program allowing for the accurate and efficient simulation of self-consistent beam dynamics. This program was used for performing the first simulations of the complete PITZ RF gun on a single PC. It can be used for future optimization and design studies.

- A systematic design study of the PITZ RF gun was carried out. The influences of all elements that break the horizontal-vertical symmetry of the gun geometry were investigated. This allowed for specifying the dedicated effects of the diagnostics doublecross section, the laser mirror and the shutter valve. It can be concluded, that only the shutter valve causes a significant asymmetry in the electron bunch distribution.

Throughout this thesis, a discussion of the problems connected with the long-term simulation of short particle bunches has been conducted and a large variety of algorithmic approaches for their solution has been presented. Nevertheless, some open questions and future research topics remain unaddressed:

- Within this thesis only conformal grid refinement has been performed. In section 3.1 is was illustrated that the application of arbitrary, non-conforming refinement leads to non-regular computational grids. Since the publication of provably stable subgridding techniques for the FIT [112], all employed numerical methods are capable of handling such grids. The interpolation and projection techniques, that have been developed, can be employed on non-regular grids without any modifications. Nevertheless, the administration of arbitrary-level non-conformal refinement can be expensive regarding the additional memory required for storing the grid connectivity and hierarchy. Also, the consumption of CPU time for performing the adaptations will largely increase. It is, hence, an open question how the overall efficiency of conforming and non-conforming refinement relate to each other. Regarding future research topics listed by EULER in [49], it can be stated that the applicability of time-adaptive grids, at least for a specific type of adaptations, has been demonstrated.

- The subject of local time steps has not been addressed. The stable time step is a local quantity, which is determined by the dimensions of the respective cell. If the dimensions vary within a large range, the application of local time steps offers a way for reducing the overall computational costs. The application of local time steps within the FIT was introduced by THOMA in [128]. In [94] PIPERNO published a method for performing symplectic local time stepping within the central flux formulation of the DGM. For both methods considerable savings in computational time were demonstrated.

- The DGM is a high order spatial discretization technique. This makes the application of high order time integration methods, as presented in [57], a natural choice. This topic was not addressed.

- The implementation of the presented automated hp-adaptive algorithm should be extended to work with three-dimensional problems.

- Adaptivity can also be introduced to the discrete charged particle dynamics. A certain number of macro particle per cell is required. Otherwise, the volatile behavior of the convective currents pollute the electromagnetic field solution with numerical noise. However, as shown in the figure 5.14a, the size of the particle bunch can undergo strong variations. A dynamic adaptation of the number of macro particles, based on their density, could considerably reduce the computational burden.

APPENDIX

A. A CHARGE CONSERVING PIC-SCHEME FOR THREE-DIMENSIONAL BEAM DYNAMICS SIMULATIONS

In [134] VILLASENOR and BUNEMAN published a particle-in-cell scheme, which is commonly employed for performing self-consistent charged particle simulations within the FDM and the FIT. The computational charges are not modeled as point charges but as charged clouds. The amount of charge, which is deposited in every primary node is calculated from the overlap of all charged clouds with the respective dual volume. This assignment is illustrated in the figure 3.33b. The convective currents are calculated from the change of the deposited charge in the nodes from one time step to another. In [11] pp. 23 the respective equations for particle motions in two-dimensional domains are given. As already presented in section 3.7.2, for this case, there are four edges and associated currents, and four nodes and associated charges (see fig. A.1). Given the change of charges, the currents can be uniquely determined by solving:

$$\begin{pmatrix}\Delta q_1\\ \Delta q_2\\ \Delta q_3\\ \Delta q_4\end{pmatrix}=\begin{pmatrix}-1 & 0 & -1 & 0\\ 1 & -1 & 0 & 0\\ 0 & 0 & 1 & -1\\ 0 & 1 & 0 & 1\end{pmatrix}\cdot\begin{pmatrix}\widehat{\widehat{j}}_1\\ \widehat{\widehat{j}}_2\\ \widehat{\widehat{j}}_3\\ \widehat{\widehat{j}}_4\end{pmatrix} \quad (A.1)$$

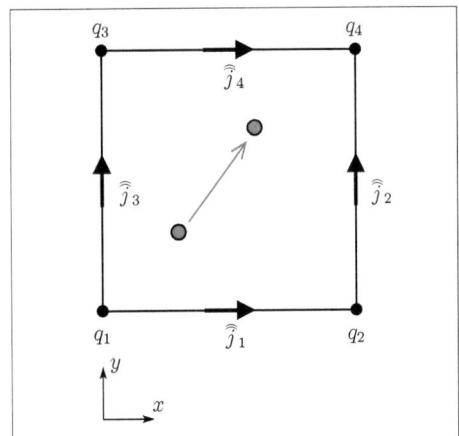

Figure A.1: PIC method in a two-dimensional cell: The charge of the particle is partially assigned to the surrounding nodes q_i according to the cloud-in-cell method depicted in figure 3.33b. The convective currents $\widehat{\widehat{j}}_i$ are calculated from the change of the assigned charges with time.

The situation is different if a three-dimensional cell is considered. Partial charges are deposited in its eight nodes which are connected by twelve edges. The problem is, thus, under-determined. In order to solve it, it has to be complemented with side conditions. In the following, a unit cell is assumed. Its lower left point coincides with the origin of the Cartesian coordinate system. A particle of charge Q travels from the location A, $(x_A, y_A, z_A)^T$, to the location B, $(x_B, y_B, z_B)^T$, within the cell. For the convective current $\widehat{\widetilde{j}}_1$, connecting the nodes located at $(0,0,0)^T$ and $(1,0,0)^T$, a reasonable side condition is given by:

$$\widehat{\widetilde{j}}_1 \sim (x_B - x_A) \cdot \frac{(1-y_A)+(1-y_B)}{2} \cdot \frac{(1-z_A)+(1-z_B)}{2} \qquad (A.2)$$

If a particle moves exactly along an edge, this condition ensures that only the discrete current associated with this edge is non-zero. Otherwise, the current amplitude scales inversely with the distance of the average particle position. All other side conditions are obtained by a cyclic exchange of variables. Further investigations show that the scheme preserves the total amount of charge, however, locally it is not charge preserving.

In the figure A.2 a charged particle travels from the position $(0.25, 0.25, 0.25)^T$ to the position $(0.75, 0.75, 0.75)^T$. Its starting position and the charges assigned to the nodes are depicted in (a) and (c), and the final position and the respective charge assignments are shown in (b) and (d). The size of the cubes scales with the amount of charge assigned to the position. Updating the charge distribution (b) with the convective currents, leads to the final distribution shown in (e). However, the distribution (e) does not agree with the reference solution (d). For the special case of a particle moving along the diagonal of a cell, the errors for all discrete charges are equal in magnitude. They correspond to $0.0312 \cdot Q$. The errors in charge are shown in (f). A surplus is indicated in gray, a deficit in black.

The application of the convective currents preserve the total amount of charge. The sum of fractional charges in (c), (d) and (e) is Q. However, errors are committed locally. In every time step, the deposited charge is recalculated from the current particle position and the particle weighting scheme. Hence, it is implicitly assumed that the charge distribution in the depicted cell at the beginning of the successive time step corresponds to (d) while actually the distribution (e) is present. Consequently, the residual charges in (f) are not propagated in the following time step but remain as static charges at their position.

Further investigations show that no residual charges stay behind if the macro particle performs movements within any plane parallel to the $x-y$-, $x-z$- or $y-z$-plane. Figure A.3 illustrates the motion of a particle from the position $(0.25, 0.25, 0.25)^T$ to the position $(0.75, 0.25, 0.75)^T$. The charge distributions (d) and (e) are identical.

The idea for a locally charge preserving PIC scheme is, thus, to split the three-dimensional particle motion into a sequence of a two-dimensional and a one-dimensional move. The application of a fixed sequence order, like always moving in the $x-y$-plane first and along the z-coordinate after, introduces a systematic error. To prevent this, the sequence can be randomized. However, the particle can also be split into three macro particles, each carrying one third of its charge. Every particle follows one of the three possible paths. This enlarges the affected number of grid currents and has a positive effect on the smoothness of the current distribution within the computational grid. The additional computational costs for performing this particle splitting are low because the position of the macro particle has to be updated only once.

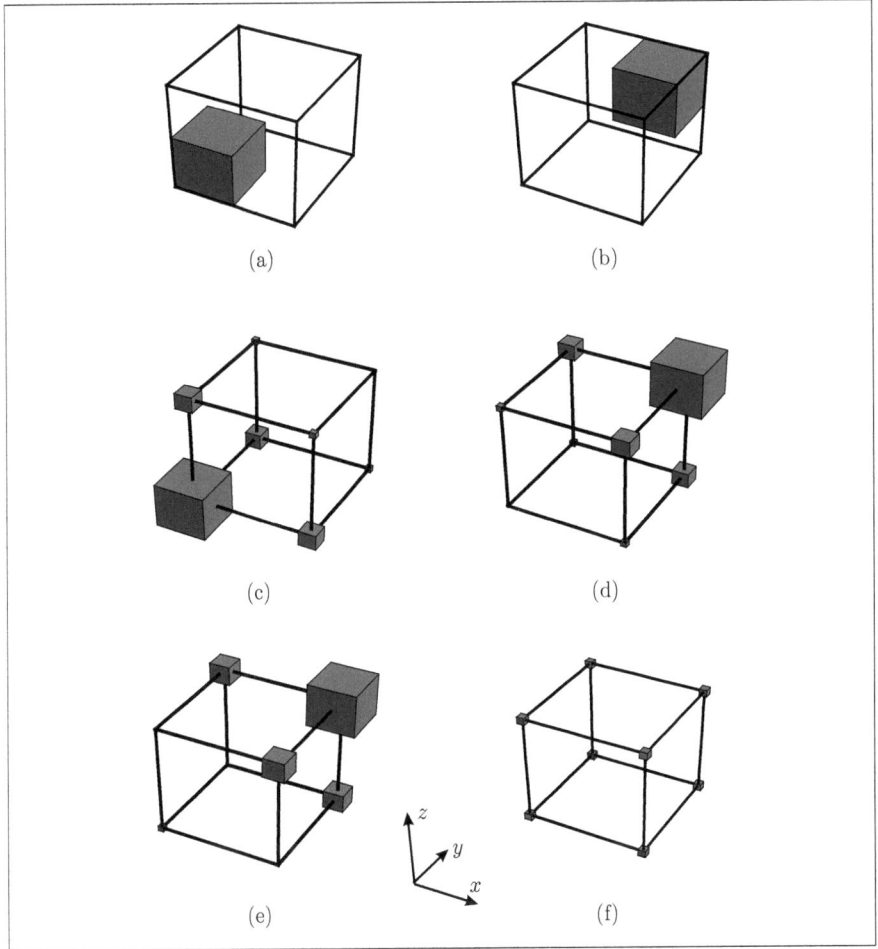

Figure A.2: PIC method in a three-dimensional cell I: A macro particle travels from the position $(0.25, 0.25, 0.25)^T$ (a) to the position $(0.75, 0.75, 0.75)^T$ (b). Performing the cloud-in-cell particle weighting yields the respective discrete charge assignments shown in (c) and (d). The size of the cubes scales with the amount of charge deposited in the node. The application of the convective currents to the distribution (c) results in the distribution (e), which is not identical with the reference distribution (d). The errors are indicated in (f). For a particle motion along the diagonal of a cell, all errors are equal in magnitude. In this case, the error is $0.03125 \cdot Q$ with Q the charge of the macro particle. A surplus of charge is indicated in gray, a deficit in black.

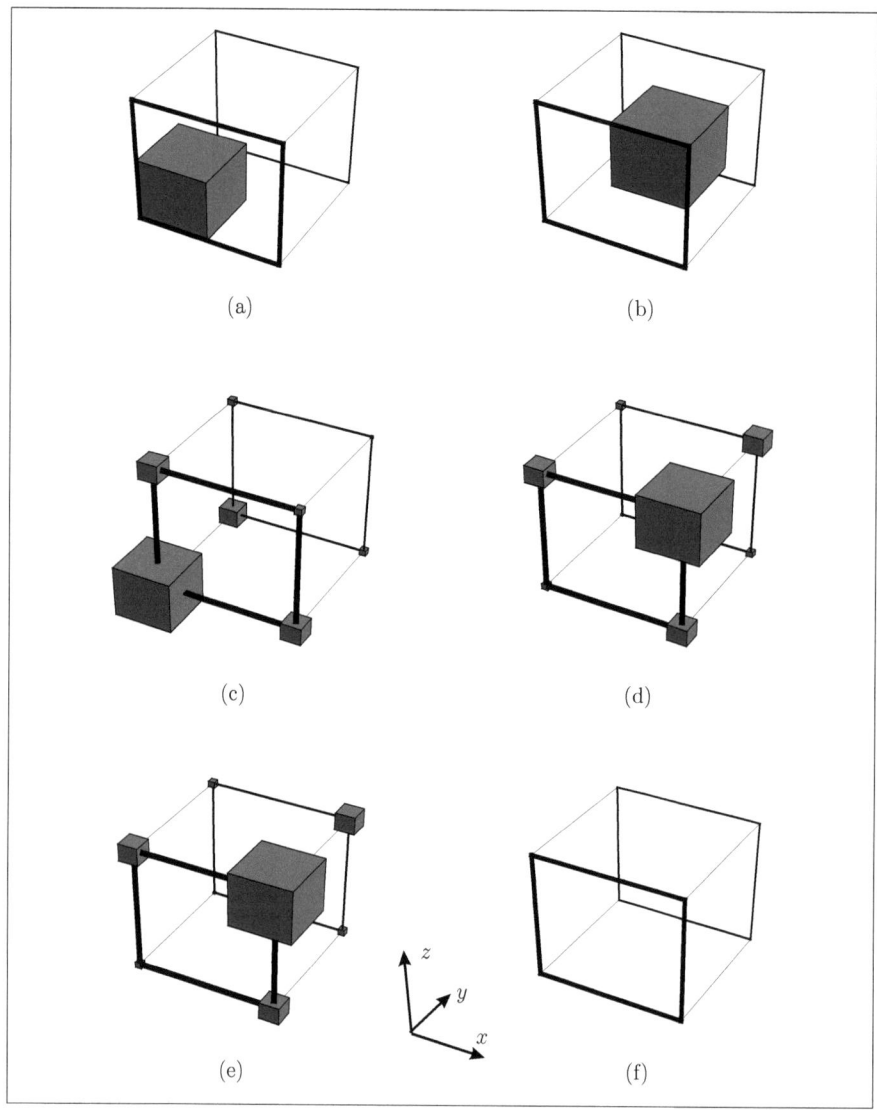

Figure A.3: PIC method in a three-dimensional cell II: Here, a macro particle travels in the $x-z$-plane from the position $(0.25, 0.25, 0.25)^\mathrm{T}$ (a) to the position $(0.75, 0.25, 0.75)^\mathrm{T}$ (b). The respective discrete charge assignments are shown in (c) and (d). For this movement the distributions shown in (d) and (e) are identical. The thickness of the edges scales with the magnitude of the associated grid current. In figure A.2, all currents are equal in magnitude.

B. PROJECTION MATRICES FOR DG-H-ADAPTATION

Projection matrices for h-refinement:

$$\Pi_l^+ = \begin{pmatrix} 1.00000 & -0.866025 & 0 & 0.330719 & 0 & -0.207289 & 0 \\ 0 & 0.500000 & -0.968246 & 0.572822 & 0.216506 & -0.359035 & -0.097578 \\ 0 & 0 & 0.250000 & -0.739510 & 0.838525 & -0.231756 & -0.377918 \\ 0 & 0 & 0 & 0.125000 & -0.496078 & 0.822653 & -0.596212 \\ 0 & 0 & 0 & 0 & 0.062500 & -0.310934 & 0.676041 \\ 0 & 0 & 0 & 0 & 0 & 0.031250 & -0.186848 \\ 0 & 0 & 0 & 0 & 0 & 0 & 0.015625 \end{pmatrix}$$

$$\Pi_r^+ = \begin{pmatrix} 1.00000 & 0.866025 & 0 & -0.330719 & 0 & 0.207289 & 0 \\ 0 & 0.500000 & 0.968246 & 0.572822 & -0.216506 & -0.359035 & 0.097578 \\ 0 & 0 & 0.250000 & 0.739510 & 0.838525 & 0.231756 & -0.377918 \\ 0 & 0 & 0 & 0.125000 & 0.496078 & 0.822653 & 0.596212 \\ 0 & 0 & 0 & 0 & 0.062500 & 0.310934 & 0.676041 \\ 0 & 0 & 0 & 0 & 0 & 0.031250 & 0.186848 \\ 0 & 0 & 0 & 0 & 0 & 0 & 0.015625 \end{pmatrix}$$

Projection matrices for h-coarsening:

$$\Pi_l^- = \begin{pmatrix} 0.500000 & 0 & 0 & 0 & 0 & 0 & 0 \\ -0.433013 & 0.250000 & 0 & 0 & 0 & 0 & 0 \\ 0 & -0.484123 & 0.125000 & 0 & 0 & 0 & 0 \\ 0.165359 & 0.286411 & -0.369755 & 0.062500 & 0 & 0 & 0 \\ 0 & 0.108253 & 0.419263 & -0.248039 & 0.031250 & 0 & 0 \\ -0.103645 & -0.179518 & -0.115878 & 0.411326 & -0.155467 & 0.015625 & 0 \\ 0 & -0.048789 & -0.188959 & -0.298106 & 0.338020 & -0.093423 & 0.007812 \end{pmatrix}$$

$$\Pi_r^- = \begin{pmatrix} 0.500000 & 0 & 0 & 0 & 0 & 0 & 0 \\ 0.433013 & 0.250000 & 0 & 0 & 0 & 0 & 0 \\ 0 & 0.484123 & 0.125000 & 0 & 0 & 0 & 0 \\ -0.165359 & 0.286411 & 0.369755 & 0.062500 & 0 & 0 & 0 \\ 0 & -0.108253 & 0.419263 & 0.248039 & 0.031250 & 0 & 0 \\ 0.103645 & -0.179518 & 0.115878 & 0.411326 & 0.155467 & 0.015625 & 0 \\ 0 & 0.048789 & -0.188959 & 0.298106 & 0.338020 & 0.093423 & 0.007812 \end{pmatrix}$$

C. EFFICIENT DYNAMIC MEMORY MANAGEMENT

Requesting memory from the computer operation system and returning it are computationally expensive operations. Since the memory requirements of a DG cell are linked to its approximation order P, every p-adaptation, principally, demands for a reallocation of its associated memory. This becomes an issue for the implementation of an hp-adaptive DG code. There, memory operations might significantly contribute to the overall simulation time. Permanent reallocations are avoided if the ab initio assigned amount of memory is large enough to hold the DoF for the prescribed maximum approximation order. In this case, however, only small numbers of cells can be used while most of the allocated memory is actually not in use. Therefore, an efficient dynamic memory management is a necessity for an adaptive DG code of practical usability. For this purpose a library meeting this criterion has been implemented. In the following its main idea is outlined.

In order to avoid frequent calls to the system memory allocation function, clusters of memory chunks are formed. The size of the chunks of each cluster match the memory demands for the DoF of one DG cell of a certain approximation order. Only clusters are allocated and freed. During library initialization an array of administrative containers for the orders of zero to P is set up. The containers hold a pointer to the list of clusters, the current number of clusters, the number of chunks per cluster, the size of one chunk and a pointer to the next free chunk. See [48] p. 13 for UML diagram representations of the containers, the clusters and the chunks.

In figure C.1, part of an exemplary computational grid and the corresponding memory map are shown. In (a) two clusters of chunks for cells of order zero and two are allocated and one cluster for cells of the order one. No cells employing third order approximations are existent and, hence, no cluster is allocated. From (a) to (b) the distribution of approximation orders has changed. Consequently, one cluster for the order of zero is freed and a new cluster for third order elements has been initialized. In practice, clusters are not immediately deleted when no element is assigned to a cell. Empty clusters are kept in memory until a threshold number of additional free chunks is exceeded. This prevents frequent reallocations of clusters in case the required number of chunks alternates around a multiple of the number of chunks per cluster.

Due to the orthogonality of the basis functions (sec. 3.4.2 p. 50) all existent coefficients remain unaltered in case of a p-enrichment. During p-reduction the higher order coefficients are deleted while the other coefficients are kept. When performing a p adaptation a memory chunk of the new approximation order is requested from the memory management system. The respective coefficients are copied to the new chunk and the old chunk is returned. No temporary memory is required. During h-adaptation new chunks are requested (h-refinement) or non-needed chunks are returned (h-coarsening).

During hp-adaptive simulations any cell can change its approximation order. This might lead

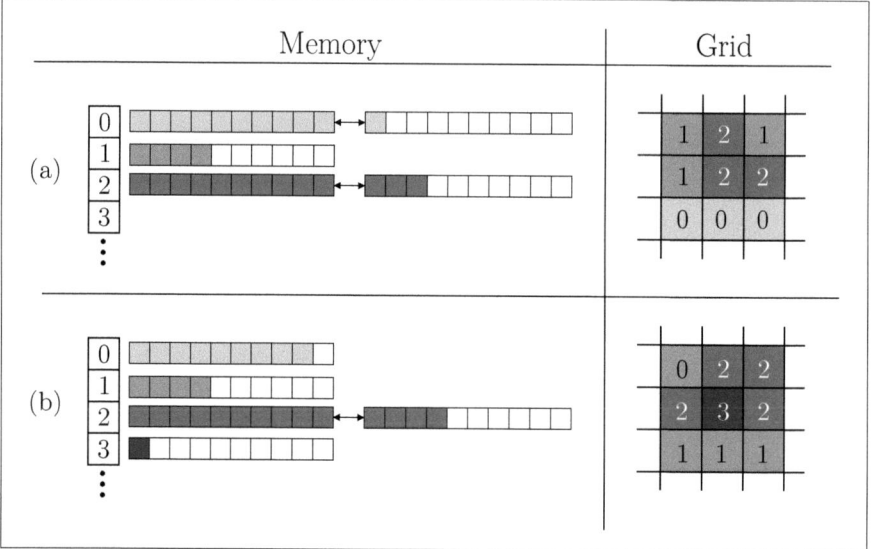

Figure C.1: Dynamic Memory Organization system: Memory map and part of an exemplary distribution of approximation orders in a DG grid. Chunks of memory that match the memory requirements of one cell of a certain approximation order are grouped in clusters. In (a) two clusters of memory chunks for the approximation order zero and two are allocated and one cluster for the order of one. For the approximation order of three no cluster is set up. In (b) the distribution of approximation orders in the grid has changed. Alongside, the memory map is adapted. One cluster for the order zero was deleted while one cluster for the order of three has been initialized. In practice, clusters are not immediately deleted when all chunks are unused. Empty clusters are kept in memory until a threshold number of additional free chunks in other clusters is exceeded.

to a distinct fragmentation of used and unused chunks within the clusters. Since clusters can be freed only if all chunks are unused, an unnecessarily high memory load can occur. This is illustrated in figure C.2. In (a) three clusters for the approximation order zero are allocated, however, two would suffice for holding the used chunks. This can be avoided by a regular check of the number of used chunks and the number of allocated clusters. If at least one cluster could be saved, a memory compression is invoked. Prior to rearranging chunks, the clusters are sorted by the number of free chunks. Then, the content of the chunks of the least occupied clusters is copied such that as many clusters as possible are completed. While sorting the clusters is computationally cheap, copying the chunk content is expensive. This sorting procedure, hence, minimizes the copy operations. In (b) the memory map after compression is shown.

The complete technical documentation of the Dynamic Memory Organization system (DynaMO) is found in [48].

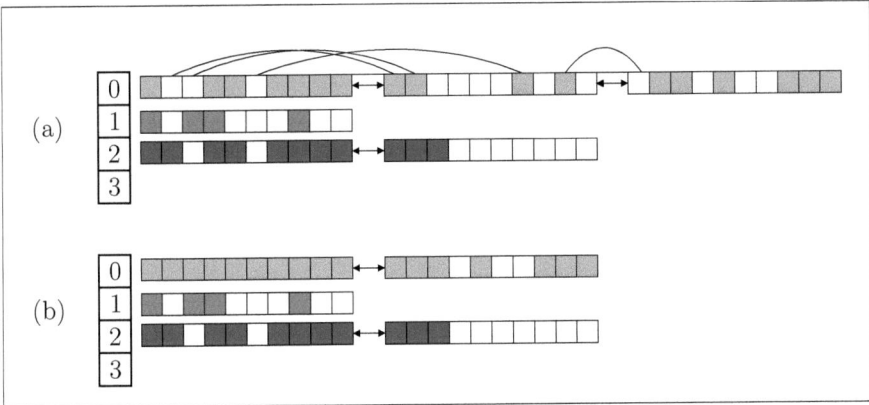

Figure C.2: Memory compression: In an hp-adaptive DG code any cell can change its approximation order. This might lead to a fragmentation of used and unused chunks within the clusters. Since only complete clusters can be freed this fragmentation increases the memory load. Therefore, a memory compression is regularly invoked. If a compression is possible, the clusters are sorted according to the smallest number of used chunks. Then, the content of the chunks of the least occupied clusters is copied such that as many clusters as possible are completed. The procedure terminates when no more cluster can be completely freed. In (a) one cluster of memory chunks for cells of the approximation order zero can be freed. Since the second cluster has the least number of used chunks, their content is copied to the other clusters. This minimizes computationally expensive copy operations. The memory map after compression is shown in (b).

D. EXEMPLARY *TAM*BCI INPUT FILE

```
#
#    ================================================================
# //                                                                 \\
# ||                    tamBCI INPUT FILE                            ||
# || --------------------------------------------------------------- ||
# ||   A program for the self-consistent and low-dispersive simulation ||
# ||   of arbitrarily shaped bunches in long accelerator structures. ||
# ||                                                                 ||
# ||   This code was developed at the Technische Universitaet Darmstadt, ||
# ||   Institut fuer Theorie Elektromagnetischer Felder (TEMF)       ||
# || --------------------------------------------------------------- ||
# \\                                                                 //
#    ================================================================
#
#
# --------------------
# GENERAL
# runnumber     : number of the run
# threads       : # of threads to be used within OpenMP regions
# priority      : process priority (1 (low) - 5 (high), 2=normal)
# particles     : use particles?
# particlemode  : tracking / consistent
# --------------------
[GENERAL]
runnumber    = 1
threads      = 4
priority     = 1
particles    = true
particlemode = consistent
#
#
# --------------------
# OUTPUT
# exportpath    : path to export VTK files to (use \\ for \)
# statistics#   : number of bunch statistics evaluations
# emitter       : write the emitter samples to VTK file?
# bunches       : export bunches to VTK file?
# astrabunches  : export bunches in ASTRA format?
```

```
# bunches#        : number of bunch files to write
# fields          : export grid, material and em-fields?
# fields#         : number of field exports
# --------------------
[OUTPUT]
exportpath    = TESLA
statistics#   = 250
emitter       = false
bunches       = true
astrabunches  = false
bunches#      = 50
fields        = true
fields#       = 50
#
#
# --------------------
# [MODEL]
# model           : load a model?
# modelname       : model file name
# shapes#         : number of shapes in the model
# shapes          : list of names of the shapes
# materials       : list of shape materials (pec, vacuum, dielectric)
# permittivity    : list of relative material permittivities
# invert          : invert shape (list)? (yes, no)
# --------------------
[MODEL]
model        = true
modelname    = FLASH.drd
shapes#      = 3
shapes       = gun; valve; mirror
materials    = pec; pec; pec
permittivity = 1.0; 1.0; 1.0
invert       = no; no; no
#
#
# --------------------
# GRID
# xsize         : xmin, xmax
# ysize         : ymin, ymax
# zsize         : zmin, zmax
# points        : x/y/z
# levels        : refinement levels x/y/z
# xfine         : min/max of fully refined part of grid x
# yfine         : min/max of fully refined part of grid y
# zfinestatic   : additional statically refined part of the grid
# zfinefactor   : refinement factor of the additional static refinement
# boundcond     : xlow, xup, ylow, yup, zlow, zup
```

```
#                   types -> electric, magnetic, open1st, open2nd
# --------------------
[GRID]
xsize          = -0.25; 0.25
ysize          = -0.5; 0.5
zsize          =  0.0; 1.0
points         =  51; 51; 201
levels         =  2; 2; 4
xfine          = -1e-3; 1e-3
yfine          = -1e-3; 1e-3
zfinestatic    = -0.001; 0.005
zfinefactor    =  0
boundcond      = electric; electric; electric; electric; electric; electric
#
#
# --------------------
# DISCRETIZATION
# method       :
#   fit           -> classical FIT with leapfrog update
#   ltfit         -> FIT with long-trans splitting (Strang-I) and leapfrog update
#                    !!! DZ <= DX, DY !!!
#   ltfv          -> long-trans splitting (Strang-I)
#                    !!! DZ <= DX, DY !!!
#                    longitudinal part treated according to Finite Volumes Method
#   dg            -> dg-fem
# fvorder      : order of FV (1 or 2)
# dgorders     : DG initial orders (x,y,z)
# dgfluxes     : flux type (sided, central)
# dgreduced    : use reduced set of basis functions (true/false)
# dgmemelements: number of lowest order elements for DynaMO library
# --------------------
[DISCRETIZATION]
method         = dg
fvorder        = 2
dgorders       = 3; 3; 2
dgfluxes       = central
dgreduced      = false
dgHP           = true
dgmemelements  = 3000
#
#
# --------------------
# EMITTER
# type         :
#   dynamic       -> calc no. of emitters according to given
#                    no. of particles per cell
#   static        -> randomly set a given no. of emitters
```

```
#   triangulated -> use triangulated surface of emission area
#   samples      : no. of emitters (per cell / globally / approx. per cell)
#   zposition    : z position of the emitter
#   rposition    : x;y position
#   radius       : emitter radius
#   species      : electrons, protons, user
#   astra        : read ASTRA bunch?
#   probability  : use every n-th sample randomly for emission
#   frequency    : emit each n-th time step (in conjunction with LT time update
#                  emission every half time step!)
# --------------------
[EMITTER]
type         = triangulated
samples      = 4
zposition    = 0.0
rposition    = 0.0; 0.0
radius       = 1e-3
species      = electrons
astra        = false
probability  = 1
frequency    = 1
#
#
# --------------------
# EMISSION
# distribr, distribt:
#   particle distribution -> gauss, rect, flattop, exp, dirac
#   paramr/paramt:
#   flattop -> mean of flat part, rise, top, fall
#   gauss   -> mean, sigma, numofsigma
#   linear  -> max
#   rect    -> mean, width
# ekin         : initial particle energy in EV
# --------------------
[EMISSION]
distribr     = rect
paramr       = 0.0; 1.0e-3
distribt     = flattop
paramt       = 11.0e-12; 2.0e-12; 18.0e-12; 2.0e-12
charge       = -1.0e-19
ekin         = 5.0e0
#
#
# --------------------
# PUSHER
#   method     : ngp, cic_locally_1D, cic_splitting
#   calls      : pusher calls in between every time update
```

```
# filter        : do filtering of excitation currents in z-direction?
# filtertype    : 3point, 5point
# factor        : distribution factor for 3point filter
# extfields     : load external fields
# --------------------
[PUSHER]
method      = cic_splitting
calls       = 1
filter      = true
filtertype  = 3point
factor      = 0.25
extfields   = true
#
#
# --------------------
# INTERPOLATION
# order         : zero, first, spline
# splinetype    : linear, akima, limited, cspline
# --------------------
[INTERPOLATION]
order       = first
splinetype  = akima
#
# ======================================================================
```

ABBREVIATIONS AND SYMBOLS

List of Acronyms

ABC	Absorbing Boundary Condition
BVP	Boundary Value Problem
CAD	Computer Aided Design
CFL	COURANT-FRIEDRICHS-LEVY
DG	Discontinuous GALERKIN
DGM	Discontinuous GALERKIN Method
DoF	Degrees of Freedom
FD	Finite Difference
FDM	Finite Difference Method
FE	Finite Element
FEM	Finite Element Method
FEL	Free Electron Laser
FI	Finite Integration
FIT	Finite Integration Technique
FLASH	Free Electron Laser in Hamburg
FV	Finite Volume
FVM	Finite Volume Method
ILC	International Linear Collider
LHC	Large Hadron Collider
LT	Longitudinal-Transverse
PBA	Perfect Boundary Approximation
PEC	Perfect Electrically Conducting
PFC	Partially Filled Cells
PITZ	Photo Injector Test facility DESY Zeuthen
RF	Radio Frequency
RMS	Root-Mean-Square
SC	Simplified Conformal
TV	Total Variation
USC	Uniformly Stable Conformal
XFEL	X-ray Free Electron Laser

List of Symbols

General

$\mathbb{R}$	real numbers
$\mathbb{N}$	natural numbers
$\mathbb{N}_0$	natural numbers including zero
$\vec{a} \cdot \vec{b}$	scalar product of vectors $\vec{a}$ and $\vec{b}$
$\vec{a} \times \vec{b}$	vector product of vectors $\vec{a}$ and $\vec{b}$
$\langle a, b \rangle$	inner product of a and b with a, b real numbers, vectors or real valued functions
δ_{ij}	KRONECKER delta
$\rho(\mathbf{A})$	spectral radius of the matrix $\mathbf{A}$
$\mathscr{F}_r$	FOURIER transform in space
$\mathscr{F}_t$	FOURIER transform in time
$\mathcal{H}$	HAMILTONIAN

Continuous Set

Ω	spatial domain with $\Omega \subseteq \mathbb{R}^3$
Ω^4	4-dimensional space-time domain
$\vec{r}$	spatial variable
x, y, z	spatial variables in CARTESIAN space
T	temporal domain with $T \in \mathbb{R}$
t	temporal variable
K, K'	inertial frames
$\mathbb{S}^N$	N-dimensional phase space
∇	gradient operator
$\nabla \cdot$	divergence operator
$\nabla \times$	curl operator
Δ	LAPLACE operator
d_t	total derivative with respect to time
∂_t	partial derivative with respect to time
$\mathrm{d}\vec{s}$	oriented infinitesimal path element
A	area
∂A	area boundary
$\mathrm{d}\vec{A}$	oriented infinitesimal area element
V	volume
∂V	volume surface
$\mathrm{d}V$	infinitesimal volume element
$\vec{n}$	normal vector
$\vec{E}$	electric field strength
$\vec{D}$	electric flux density

List of Symbols

$\vec{H}$	magnetic field strength
$\vec{B}$	magnetic flux density
ρ	electric charge density
q	electric charge
$\vec{J}$	electric current density
$\vec{J}_\kappa$	conductive current density
$\vec{J}_c$	convective current density
$\vec{J}_e$	externally impressed current density
$\vec{I}$	electric current
W	electromagnetic energy
w	electromagnetic energy density
$\vec{S}$	POYNTING vector
$\vec{A}$	magnetic vector potential
ϕ	electric scalar potential
$\vec{k}$	wave vector
ω	angular frequency
f	frequency
$\vec{v}$	velocity vector
c	speed of light
Z	space impedance
Y	space admittance
ϵ	permittivity
μ	permeability
κ	conductivity
$\vec{F}$	force vector
m	mass
$\vec{p}$	kinetic momentum
$\mathcal{P}$	generalized coordinate
$\mathcal{Q}$	generalized canonical momentum conjugate to $\mathcal{P}$
$\mathcal{F}$	flux function

Statistics

ϱ	phase space density
$\hat{\mu}^{(k,m)}$	m-k-th order raw moment
$\mu^{(k,m)}$	m-k-th order centered moment
μ	mean value
σ	standard deviation
σ^2	variance
$\mu^{(1,1)}$	covariance
ε	emittance
$\mathbf{V}$	covariance matrix

Beam Dynamics

β	relative velocity factor
γ	LORENTZ factor
B	beam brightness
x	horizontal coordinate
y	vertical coordinate
s	curvilinear particle coordinate

Discrete Set

$\mathcal{G}$	computational grid
$\tilde{\mathcal{G}}$	computational grid dual to $\mathcal{G}$
$\mathcal{G}_i$	i-th cell of the computational grid
$\Omega_{\mathcal{G}}$	computational domain
$\mathbb{P}$	set of grid points
$\mathbb{E}$	set of grid edges
$\mathbb{A}$	set of grid faces
$\mathbb{V}$	set of grid volumes
u	CARTESIAN coordinate $u \in \{x,y,z\}$
Δu	grid step size in the direction of u
$\mathbf{A}$	system matrix
λ_A	eigenvalues of the system matrix
$\mathbf{T}$	iteration matrix
λ_T	eigenvalues of the iteration matrix
k_u	discrete wave number in direction of u
β_u	phase advance per cell in direction of u
ν	COURANT-FRIEDRICHS-LEVY (CFL) number
$\widehat{e}$	FIT electric grid voltage
$\widehat{h}$	FIT magnetic grid voltage
$\widehat{\widehat{d}}$	FIT electric grid flux
$\widehat{\widehat{b}}$	FIT magnetic grid flux
$\widehat{\widehat{j}}$	FIT grid current
q	FIT grid node charge
M_ϵ	FIT permittivity parameter
M_μ	FIT permeability parameter
M_κ	FIT conductivity parameter
$\widehat{\mathbf{e}}$	FIT vector of electric grid voltages
$\widehat{\mathbf{h}}$	FIT vector of magnetic grid voltages
$\widehat{\widehat{\mathbf{d}}}$	FIT vector of electric grid fluxes
$\widehat{\widehat{\mathbf{b}}}$	FIT vector of magnetic grid fluxes
$\widehat{\widehat{\mathbf{j}}}$	FIT vector of grid currents
$\mathbf{q}$	FIT vector of grid node charges
$\mathbf{M}_\epsilon$	FIT permittivity matrix
$\mathbf{M}_\mu$	FIT permeability matrix

List of Symbols

$\mathbf{M}_\kappa$	FIT conductivity matrix
$\mathbf{P}_u$	FIT primary derivative matrix (derivative in u-direction)
$\tilde{\mathbf{P}}_u$	FIT dual derivative matrix (derivative in u-direction)
$\mathbf{S}$	FIT primary divergence matrix
$\tilde{\mathbf{S}}$	FIT dual divergence matrix
$\mathbf{C}$	FIT primary curl matrix
$\tilde{\mathbf{C}}$	FIT dual curl matrix
e	FVM electric field variable
h	FVM magnetic field variable
d	FVM electric flux density variable
b	FVM magnetic flux density variable
E^+	FVM electric field value on positive side of cell boundary
E^-	FVM electric field value on negative side of cell boundary
E^*	FVM electric field value at cell boundary
H^+	FVM magnetic field value on positive side of cell boundary
H^-	FVM magnetic field value on negative side of cell boundary
H^*	FVM magnetic field value at cell boundary
$\vec{f}$	FVM vector of numerical fluxes
s	FVM slope
$\mathbf{U}$	FVM upwind (block)matrix
$\mathbf{T}$	FI-FV variable transformation matrix
$\bar{E}$	DGM approximation of the electric field
$\bar{H}$	DGM approximation of the magnetic field
$(e)^p$	DGM electric field DoF of order p
$(h)^p$	DGM magnetic field DoF of order p
$\mathbf{e}$	DGM vector of electric field DoF
$\mathbf{h}$	DGM vector of magnetic field DoF
P	DGM approximation order
φ^p	DGM basis function of order p
$\{\varphi^p\}$	DGM set of basis functions
$\mathcal{V}^P$	DGM space spanned by set of basis functions
$\mathbf{\Phi}$	DGM vector of basis functions
Π^p	DGM projection operator to basis function of order p
ε	DGM approximation error
M	DGM mass term
S	DGM stiffness term
$\mathbf{M}$	DGM mass matrix
$\mathbf{S}$	DGM stiffness matrix
Δt	time increment
n	time step number
$\mathcal{T}$	time integration operator
$\mathbf{T}$	iteration matrix

$\mathcal{T}$	vector of truncation errors
$\mathcal{E}$	vector of errors of the discrete field solution
$\bar{\mathcal{E}}$	vector of error estimates
R	residual

Refinement Set

L	refinement level
$\mathcal{I}$	general interpolation operator
S	interpolating spline function
$\overline{w}$	interpolated space-continuous energy density
$\overline{\mathcal{W}}$	integrated space-continuous energy density
P'	FIT new primary grid node
$\widetilde{P}'$	FIT new dual grid node
$\widehat{e}\,'$	FIT interpolated electric grid voltage
$\widehat{h}\,'$	FIT interpolated magnetic grid voltage
$\bar{E}^+$	DGM h-refined approximation of the electric field
$\bar{H}^+$	DGM h-refined approximation of the magnetic field
$(e)_\mathrm{l}^l$	DGM left h-refined electric field DoF of order l
$(e)_\mathrm{r}^r$	DGM right h-refined electric field DoF of order r
$(h)_\mathrm{l}^l$	DGM left h-refined magnetic field DoF of order l
$(h)_\mathrm{r}^r$	DGM right h-refined magnetic field DoF of order r
$(\mathbf{e})_\mathrm{l}$	DGM vector of left h-refined electric field DoF
$(\mathbf{e})_\mathrm{r}$	DGM vector of right h-refined electric field DoF
$(\mathbf{h})_\mathrm{l}$	DGM vector of left h-refined magnetic field DoF
$(\mathbf{h})_\mathrm{r}$	DGM vector of right h-refined magnetic field DoF
L	DGM left h-refined element approximation order
R	DGM right h-refined element approximation order
φ_l^l	DGM left h-refined element basis function of order l
φ_r^r	DGM right h-refined element basis function of order r
$(\Pi_\mathrm{l}^{l,p})^+$	DGM projection operator to left h-refined element of order l
$(\Pi_\mathrm{r}^{r,p})^+$	DGM projection operator to right h-refined element of order r
Π_l^+	DGM projection matrix to left h-refined element
Π_r^+	DGM projection matrix to right h-refined element
$(\Pi_\mathrm{l}^{p,l})^-$	DGM projection operator from left hand side element to h-coarsened element of order p
$(\Pi_\mathrm{r}^{p,r})^-$	DGM projection operator from right hand side element to h-coarsened element of order p
Π_l^-	DGM projection matrix from left hand side elements to h-coarsened element
Π_r^-	DGM projection matrix from right hand side elements to h-coarsened element
η	resolution level of a DG cell

List of Symbols

η^* .... designated resolution level
$\hat{\eta}$ .... preset maximum resolution level
d .... progressivity parameter of resolution level mapping

LIST OF FIGURES

1.1	FLASH injector section	4
1.2	FLASH injector and adaptively refined computational grid	6
2.1	Illustration of Ampère's and Faraday's law	11
2.2	Phase space diagram	19
2.3	Illustration of the connection between the state of a beam and its emittance	22
3.1	Discretization of continuous space	24
3.2	Hexahedral grid cell	25
3.3	Regular grid and conformal refinement	25
3.4	Organization of the grid refinement in a tree structure	26
3.5	Topology of refined grids at two instants in time and their tree representation	27
3.6	Staircase material fillings	27
3.7	Time-adaptive material discretization	28
3.8	Illustration of Ampère's and Faraday's law discretized by FIT	30
3.9	Staggered pair of dual orthogonal grids	31
3.10	FIT on dual orthogonal grids	32
3.11	Dispersion diagram of the FIT	36
3.12	Dispersion diagrams of FIT and FV	38
3.13	Illustration of the FI-FV Method in space	39
3.14	FI-FV grid setup and coupling of DoF	44
3.15	Grid dispersion and dissipation diagrams of the Godunov, Lax-Wendroff and Fromm method	46
3.16	Scaled Legendre polynomials	50
3.17	Population of the DG curl matrix	53
3.18	Dispersion diagrams of the DG method I	56
3.19	Dispersion diagrams of the DG method II	57

3.20	Eigenvalues of the FIT and DG system matrix and iteration matrix in the complex plane	60
3.21	Eigenvalues of the FI-FV hybrid system matrix and iteration matrices in the complex plane	62
3.22	Eigenvalues of the FI-FV hybrid system matrix and the transverse and longitudinal system matrix	65
3.23	Eigenvalues of the FI-FV longitudinal system matrix and iteration matrix in the complex plane	65
3.24	Dispersion error of the FIT and FVMs	69
3.25	Dispersion error of the DGM	69
3.26	Stable CFL numbers for the DGM	71
3.27	Dispersion diagrams of the fully discrete FIT and DGM	72
3.28	Dispersion errors of the fully discrete FIT and DGM	73
3.29	Amplification and angular frequency of the GODUNOV, LAX-WENDROFF and FROMM scheme	74
3.30	Amplitude and frequency error of the GODUNOV, LAX-WENDROFF and FROMM scheme	75
3.31	Dispersion diagram of the fully discrete FIT (1D)	76
3.32	Eigenvalues of the iteration matrix using GODUNOV and STRANG splitting . . .	76
3.33	Particle weighting schemes (PIC)	80
3.34	FIT charge assignment and grid currents for different particle weighting schemes	83
3.35	PIC method in a two-dimensional cell	84
3.36	Field interpolation to the particle position (PIC)	84
3.37	Comparison of particle pushers	85
4.1	Grid adaptation within the FIT: Allocation of electric grid voltages	98
4.2	Grid adaptation within the FIT: Allocation of magnetic grid voltages	99
4.3	Grid adaptation within the FIT: Interpolation of electric grid voltages	100
4.4	Grid adaptation within the FIT: Interpolation of magnetic grid voltages	100
4.5	Grid adaptation within the FIT: Comparison of spline interpolations	101
4.6	Grid adaptation within the FIT: Setup and evaluation of the interpolating spline	101
4.7	Grid adaptation within the FIT: Determination of continuous sampled quantities	102
4.8	Grid adaptation within the FIT: Adaptation of the particle cloud size	103
4.9	Grid adaptation within the DGM: Basis functions and projection for h-refinement	108
4.10	Grid adaptation within the DGM: Projection based grid refinement and coarsening	110

List of Figures

4.11 Grid adaptation within the DGM: Dependence of the convergence rate on the smoothness of the function 115

4.12 Grid adaptation within the DGM: Smoothness indicator 117

4.13 Grid adaptation within the DGM: Illustration of the strategy for resolution adaptation 119

5.1 Verification of the grid adaptation functionality 122

5.2 Verification of the interpolation procedures 123

5.3 Cylindrical wave simulated with FIT 126

5.4 Cylindrical wave simulated with FI-FV scheme 127

5.5 Cylindrical wave simulated with FI-FV scheme and reduced time step 128

5.6 Cylindrical wave simulated with LT-FIT and reduced time step 129

5.7 Dispersion of the DGM for various approximation orders 130

5.8 Dispersion of FIT and LT-FIT 131

5.9 Numerical phase velocity and noise level vs. refinement level 131

5.10 Simulation of a GAUSSIAN and trapezoidal wave packet using the automated hp-adaptive algorithm 132

5.11 Longitudinal component of the electric field excited by a GAUSSIAN bunch traveling in a tube 134

5.12 Computed RMS bunch length vs. longitudinal grid step size 139

5.13 Computed RMS bunch width and length vs. number of emitters per cell 140

5.14 Evolution of bunch parameters in the PITZ RF gun 141

5.15 Models and results of the PITZ gun design study 142

A.1 PIC method in a two-dimensional cell 149

A.2 PIC method in a three-dimensional cell I 151

A.3 PIC method in a three-dimensional cell II 152

C.1 Dynamic memory organization system 156

C.2 Memory compression 157

BIBLIOGRAPHY

[1] W. ACKERMANN, E. SALDIN, E. SCHNEIDMILLER, AND M. YURKOV ET AL., *Operation of a free electron laser in the wavelength range from the extreme ultraviolet to the water window*, Nature Photonics, 1 (2007), pp. 336–342. Cited on page 3

[2] S. ADJERID, K. DEVINE, J. FLAHERTY, AND L. KRIVODONOVA, *A posteriori error estimation for discontinuous galerkin solutions of hyperbolic problems*, Comput. Methods. Appl. Mech. Eng., 191 (2002), pp. 1097–1112. Cited on page 113

[3] M. AINSWORTH, *Dispersive and dissipative behaviour of high order discontinuous galerkin finite element methods*, Journal of Computational Physics, 198 (2004), pp. 106–130. Cited on pages 53, 54, and 89

[4] M. AINSWORTH AND B. SENIOR, *Aspects of an adaptive hp-finite element method: Adaptive strategy, conforming approximation and efficient solvers*, Comput. Methods. Appl. Mech. Eng., 150 (1997), pp. 65–87. Cited on page 116

[5] H. AKIMA, *A new method of interpolation and smooth curve fitting based on local procedures*, Jour. ACM, 17 (1970), pp. 589–602. Cited on page 93

[6] M. ALTARELLI, R. BRINKMANN, M. CHERGUI, W. DECKING, B. DOBSON, S. DÜSTERER, G. GRÜBEL, W. GRAEFF, H. GRAAFSMA, J. HAJDU, J. MARANGOS, J. PFLÜGER, H. REDLIN, D. RILEY, I. ROBINSON, J. ROSSBACH, A. SCHWARZ, K. TIEDTKE, T. TSCHENTSCHER, I. VARTANIANTS, H. WABNITZ, H. WEISE, R. WICHMANN, K. WITTE, A. WOLF, M. WULFF, AND M. YURKOV, *The european x-ray free-electron laser - technical design report*, (2007). Cited on page 3

[7] A. M. AMPÈRE, *Théorie Mathématique des Phénomènes Électro-Dynamiques*, Libraire Scientifique Albert Blanchard, 1827. Cited on page 9

[8] S. BADEN, N. CHRISOCHOIDES, D. GANNON, AND M. NORMAN, eds., *Structured Adaptive Mesh Refinement (SAMR) Grid Methods*, Springer, 2000. Cited on page 89

[9] J. BARTH, *Adaptive Mesh Refinement - Theory and Applications*, Springer, 2005, ch. A posteriori error estimation and adaptivity for FV and FE methods, pp. 183–202. Cited on page 68

[10] J. BÜCHNER, C. DUM, AND M. SCHOLER, eds., *Space Plasma Simulation*, Springer, 2003. Cited on page 79

[11] U. BECKER, *Numerische Berechnung elektromagnetischer Felder in Wechselwirkung mit frei beweglichen Ladungen*, PhD thesis, TU Darmstadt, 1997. Cited on pages 82 and 149

[12] M. BERGER AND P. COLELLA, *Local adaptive mesh refinement for shock hydrodynamics*, Journal of Computational Physics, 82 (1989), pp. 64–84. Cited on page 88

[13] M. BERGER AND R. J. LEVEQUE, *Adaptive mesh refinement using wave-propagation algorithms for hyperbolic systems*, SIAM J. Numerical Analysis, 35 (1997), pp. 2298–2316. Cited on pages 5, 88, and 143

[14] N. BESSE, F. FILBERT, M. GUTNIC, I. PAUN, AND E. SONNENDRÜCKER, *An adaptive numerical method for the Vlasov equation based on a multiresolution analysis*, Numerical Mathematics and Advanced Applications ENUMATH 2001, Springer, 2001, pp. 437–446. Cited on page 78

[15] W. BIALOWONS, M. BIELER, H.-D. BREMER, F.-J. DECKER, H.-C. LEWIN, P. SCHÜTT, G.-A. VOSS, R. WANZENBERG, AND T. WEILAND, *Computer simulations of the wake field transformer experiment at desy*, in Proc. of the 1st European Particle Accelerator Conference (EPAC), vol. 1, 1988, pp. 902–904. Cited on page 29

[16] C. K. BIRDSALL, *Particle-in-cell charged-particle simulations, plus monte carlo collisions with neutral atoms, pic-mcc*, IEEE Trans. Plasma Science, 19 (1991), pp. 65–85. Cited on page 79

[17] ———, *Plasma physics via computer simulation*, Institute of Physics, London, 1991. Cited on page 81

[18] P. BONNET, X. FERRIERES, B. MICHIELSEN, P. KLOTZ, AND J. ROUMIGUIÈRES, *Time Domain Electromagnetics*, Academic Press, 1999, ch. Finite-Volume time domain method, pp. 307–367. Cited on page 38

[19] J. BORIS, *Relativistic plasma simulation – optimization of a hybrid code*, in Proc. 4th Conf. Numer. Simulation of Plasmas, 1970. Cited on page 82

[20] A. I. BORISENKO AND I. E. TARAPOV, *Vector and Tensor Analysis with Applications*, Dover Publications, 1979. Cited on page 10

[21] A. BOSSAVIT, *Computational electromagnetism*, Academic Press, 1997. Cited on pages 5, 47, and 51

[22] A. BOSSAVIT AND L. KETTUNEN, *Yee-like schemes on staggered cellular grids: A synthesis between fit and fem approaches*, IEEE Trans., MAG, 36 (2000), pp. 861–867. Cited on page 97

[23] J. BRAU, Y. OKADA, AND N. WALKER, *Ilc reference design report - executive summary*, (2007). Cited on page 3

[24] O. BRÜNING ET AL., *Lhc design report*, CERN-2004-003, (2004). Cited on page 3

[25] I. BRONSTEIN, K. SEMENDYAYEV, G. MUSIOL, AND H. MUEHLIG, *Handbook of Mathematics*, Springer. Cited on page 17

[26] J. BUON, *Beam phase space and emittance*, in CAS - CERN Accelerator School: 5th General accelerator physics course, S. Turner, ed., vol. I, 1992. Cited on pages 16, 18, and 20

[27] A. CANDEL, A. KABEL, L. LEE, Z. LI, C. LIMBORG, C. NG, E. PRUDENCIO, R. SCHUSSMAN, R. UPLENCHWAR, AND K. KO, *Parallel finite element particle-in-cell code for simulations of space-charge dominated beam-cavity interactions*, in Proc. Particle Acc. Conf. (PAC), Albuquerque, USA, 2006. Cited on page 88

[28] N. CANOUET, L. FEZOUI, AND S. PIPERNO, *Discontinuous galerkin time-domain solution of maxwell's equations on locally-refined nonconforming cartesian grids*, COMPEL, 24 (2005), pp. 1381–1401. Cited on pages 5, 89, and 143

[29] B. CARLSTON, *New photoelectric injector design for the los alamos national laboratory xuv fel accelerator*, NIM A, 285 (1989), pp. 313–319. Cited on page 138

[30] R. CEE, M. KRASSILNIKOV, S. SETZER, T. WEILAND, AND A. NOVOKHATSKI, *Detailed numerical studies of space charge effects in an FEL RF gun*, Nucl. Inst. and M. Phys. A, 483 (2002), pp. 321–325. Cited on page 138

[31] P. CHANNELL, *The moment approach to charged particle beam dynamics*, IEEE Trans. Nucl. Sci., 30 (1983), p. 2608. Cited on page 78

[32] A. W. CHAO AND M. TIGNER, eds., *Handbook of Accelerator Physics and Engineering*, World Scientific, 1999. Cited on page 19

[33] M. CLEMENS, R. SCHUHMANN, AND T. WEILAND, *Algebraic properties and conservation laws in the discrete electromagnetism*, Frequenz, 53 (1999), pp. 219–225. Cited on pages 33, 34, and 59

[34] B. COCKBURN, *Discontinuous galerkin methods*, Z. Angewandte Mathematische Mechanik (ZAMM), 83 (2003), pp. 731–754. Cited on page 114

[35] ———, *A simple introduction to error estimation for nonlinear hyperbolic conservation laws*, School of Mathematics, University of Minnesota, (2003). Cited on page 113

[36] B. COCKBURN AND C.-W. SHU, *Tvb runge-kutta local projection discontinuous galerkin finite element method for conservation laws*, Mathematics of Computation, I-V (1989-1998). Cited on page 89

[37] A. COHEN, S. KABER, S. MÜLLER, AND M. POSTEL, *Fully adaptive multiresolution finite volume schemes for conservation laws*, Mathematics of Computation, 72 (2000), pp. 183–225. Cited on pages 88 and 113

[38] G. COHEN, *Higher-Order Numerical Methods for Transient Wave Equations*, Springer, 2002. Cited on pages 13, 47, and 67

[39] P. COLELLA AND E. G. PUCKETT, *Modern numerical methods for fluid flow*, UC Berkeley and UC Davis, (1998). Cited on page 35

[40] CST GMBH, Bad Nauheimer Str. 19, 64289 Darmstadt, Germany. Cited on pages 27, 88, and 138

[41] C. DE BOOR, *A Practical Guide to Splines*, Springer-Verlag, 2001. Cited on page 93

[42] M. DE LOOS AND S. VAN DER GEER, *General particle tracer: A new 3d code for accelerator and beamline design*, in Proc. 6th Europ Particle Accelerator Conference (EPAC), 1996. Cited on page 79

[43] M. DEHLER, A. CANDEL, AND E. GJONAJ, *Full scale simulation of a field-emitter arrays based electron source for free-electron lasers*, J. Vac. Sci. Technol. B, 24 (2006), pp. 892–897. Cited on pages 82 and 89

[44] M. DEHLER AND T. WEILAND, *Calculating wake potentials for ultra short bunches*, in AIP Conference Proceedings, vol. 297, 1993, pp. 123–130. Cited on page 88

[45] L. DEMKOWICZ, *Computing with hp-Adaptive Finite Elements*, vol. I and II, Chapman & Hall/CRC, 2007/2008. Cited on pages 5, 88, 109, 114, 120, 143, and 144

[46] S. DEY AND R. MITTRA, *A locally conformal finite-difference time-domain(fdtd) algorithm for modeling three-dimensional perfectly conducting objects*, IEEE Microwave Guided Wave Lett., (1997), p. 273. Cited on page 27

[47] J. E. DRUMMOND, *Plasma physics*, McGraw-Hill, 1961. Cited on page 78

[48] DYNAMO, *A library for the efficient dynamic memory management within an adaptive dg code*. Technical documentation: http://www.temf.de/downloads/DynaMO.pdf. Cited on pages 155 and 156

[49] T. EULER, *Consistent Discretization of Maxwell's Equations on Polyhedral Grids*, PhD thesis, TU Darmstadt, 2007. Cited on pages 95, 97, and 146

[50] H. FAHS, S. LANTERI, AND F. RAPETTI, *A hp-like discontinuous galerkin method for solving the 2d time-domain maxwell's equations on non-conforming locally refined triangular meshes*, tech. report, RR-6162, INRIA, http://hal.inria.fr/ inria-00140783/fr, 2007. Cited on page 89

[51] M. FARADAY, *Experimental Researches in Electricity*, Bernard Quaritch, London, 1855. Cited on page 9

[52] K. FLÖTTMANN, S. LIDIA, AND P. PIOT, *Recent improvements to the astra particle tracking code*, in Proc. of the Particle Accelerator Conference (PAC), 2003. Cited on page 79

[53] K. FLÖTTMANN AND F. STEPHAN, *Proposal for the bmbf "rf photoinjectors as sources for electron bunches of extremely short length and small emittance"*, (1999). Cited on page 138

[54] D. GAITONDE, J. SHANG, AND J. YOUNG, *Practical aspects of higher-order numerical schemes for wave propagation phenomena*, Int. J. Num. Methods in Engineering, 45 (1999), pp. 1849–1869. Cited on page 81

[55] E. GJONAJ, *Dispersion and stability analysis of the discontinuous galerkin method for maxwell equations*, Advances in Electromagnetic Research, Kleinwalsertal, Austria, (2006). Cited on pages 54 and 68

[56] E. GJONAJ, X. DONG, R. HAMPEL, M. KÄRKKÄINEN, T. LAU, W. MÜLLER, AND T. WEILAND, *Large scale parallel wake field computations for 3d accelerator structures with the pbci code*, in Proc. of the 9th Int. Comp. Acc. Phys. Conf. (ICAP), 2006. Cited on page 141

[57] E. GJONAJ, T. LAU, S. SCHNEPP, F. WOLFHEIMER, AND T. WEILAND, *Accurate modelling of charged particle beams in linear accelerators*, New Journal of Physics, 8 (2006), pp. 1–21. Cited on pages 49, 52, 59, 63, 82, 88, 89, 138, and 146

[58] E. GJONAJ, T. LAU, AND T. WEILAND, *Conservation properties of the discontinuous galerkin method for maxwell equations*, in Proc. Int. Conf. Electromagnetics in Advanced Applications (ICEAA), 2007. Cited on page 53

[59] S. GODUNOV, *A finite difference method for the numerical computation and discontinuous solutions of the equations of fluid dynamics*, Mat. Sb., 47 (1959), pp. 271–306. Cited on page 63

[60] M. GROTE, A. SCHNEEBELI, AND D. SCHÖTZAU, *Interior penalty discontinuous galerkin method for maxwell's equations: Energy norm error estimates*, J. Comp. Appl. Math., 204 (2007), pp. 375–386. Cited on page 114

[61] E. HAIRER AND G. WANNER, *Solving ordinary differential equations*, vol. 1, Springer, 2 ed., 1993. Cited on page 61

[62] R. HAMPEL, *A Directionally Dispersion-free Algorithm for the Calculation of Wake Potentials*, PhD thesis, TU Darmstadt, 2008. Cited on page 14

[63] A. HARTEN, *High resolution for hyperbolic conservation laws*, J. Comp. Phys., 49 (1983), pp. 357–393. Cited on page 133

[64] J. S. HESTHAVEN AND G. JACOBS, *High-order nodal discontinuous galerkin particle-in-cell method on unstructured grids*, Journal of Computational Physics, 214 (2005), pp. 96–121. Cited on pages 81 and 82

[65] J. S. HESTHAVEN AND T. WARBURTON, *Nodal Discontinuous Galerkin Methods*, Springer, 2008. Cited on pages 5 and 47

[66] R. HOCKNEY AND J. EASTWOOD, *Computer Simulation Using Particles*, IOP Publishing, 1988. Cited on page 82

[67] P. HOUSTON AND E. SÜLI, *Error Estimation and Adaptive Discretization Methods in Computational Fuild Dynamics*, Springer, 2003, ch. Adaptive Finite Element Approximation of Hyperbolic Problems, pp. 269–344. Cited on page 113

[68] F. HU AND H. ATKINS, *Eigensolution analysis of the discontinuous galerkin method with non-uniform grids, part i: One space dimension*, tech. report, ICASE, 2001. Cited on pages 53, 68, and 89

[69] F. HU, M. HUSSAINI, AND P. RASETARINERA, *An analysis of the discontinuous galerkin method for wave propagation problems*, J. Comp. Phys., 151 (1999), pp. 921–946. Cited on pages 53 and 89

[70] Z. HUANG AND K.-J. KIM, *Review of x-ray free-electron laser theory*, PRST-AB, 10 (2007). Cited on page 138

[71] J. D. JACKSON, *Classical Electrodynamics*, Wiley & Sons, third ed., 1999. Cited on pages 12, 13, 14, and 16

[72] A. KOLMOGOROV AND S. FOMIN, *Functional Analysis*, vol. 2 (Measure, Lebesgue Integral, Hilbert Space), Graylock Press, 1961. Cited on page 48

[73] M. KRASSILNIKOV, A. NOVOKHATSKI, B. SCHILLINGER, S. SETZER, T. WEILAND, W. KOCH, AND P. CASTRO, *V-code beam dynamics simulation*, in Proc. ICAP, 2000. Cited on page 78

[74] E. KUBATKO, J. WESTERINK, AND C. DAWSON, *hp discontinuous galerkin methods for advection dominated problems in shallow water flow*, Comput. Methods. Appl. Mech. Eng., 196 (2006), pp. 437–451. Cited on pages 89 and 114

[75] P. KUS, P. SOLIN, AND I. DOLEZEL, *Solution of 3d singular electrostatics problems using adaptive hp-fem*, COMPEL, 27/3 (2008). Cited on page 88

[76] L. D. LANDAU AND E. M. LIFSHITZ, *Course of Theoretical Physics*, vol. I Mechanics, Pergamon Press, 1960. Cited on page 16

[77] C. B. LANEY, *Computational Gasdynamics*, Cambridge University Press, 1998. Cited on pages 39, 40, 49, 64, 66, 70, 94, and 133

[78] J. LANG, *Adaptive Multilevel Solution of Nonlinear Parabolic PDE Systems*, Springer, 2001. Cited on page 88

[79] T. LAU, *Numerische Methoden zur Simulation teilchengenerierter elektromagnetischer Felder in der Beschleunigerphysik*, PhD thesis, TU Darmstadt, 2006. Cited on pages 41, 42, 81, and 82

[80] T. LAU, E. GJONAJ, AND T. WEILAND, *Time integration methods for particle beam simulations with the finite integration theory*, Zeitschrift für Telekommunikation (FREQUENZ), 59 (2005), pp. 210–219. Cited on pages 37, 62, and 77

[81] J. LE DUFF, *Longitudinal beam dynamics in circular accelerators*, in CAS - CERN Accelerator School: 5th General accelerator physics course, S. Turner, ed., vol. I, 1992. Cited on page 16

[82] R. LEIS, *Initial Boundary Value Problems in Mathematical Physics*, Wiley & Sons, 1986. Cited on page 13

[83] R. J. LEVEQUE, *Numerical Methods for Conservation Laws*, Birkhäuser, 1990. Cited on pages 5, 37, 39, 49, and 94

[84] ———, *Finite Volume Methods for Hyperbolic Problems*, Cambridge Texts in Applied Mathematics, 2002. Cited on page 37

[85] J. MAXWELL, *On physical lines of force*, Philosophical Magazine and Journal of Science, (1861). Cited on page 9

[86] ——, *A dynamical theory of the electromagnetic field*, Philosophical Transactions of the Royal Society of London, 155 (1865), pp. 459–512.

[87] ——, *A Treatise on Electricity and Magnetism*, Oxford University Press, first ed., 1873. Cited on page 9

[88] F. MAYER, R. SCHUHMANN, AND T. WEILAND, *Flexible subgrids in FDTD calculations*, IEEE International Symposium on Antennas and Propagation, 3 (2002), pp. 252–255. Cited on page 87

[89] J. MELENK, *hp-Finite Element Methods for Singular Perturbations*, Springer, 2002. Cited on page 113

[90] K. W. MORTON AND D. F. MAYERS, *Numerical Solution of Partial Differential Equations*, Cambridge University Press, 2005. Cited on page 35

[91] C.-D. MUNZ, R. SCHNEIDER, AND U. VOSS, *A finite-volume method for the maxwell equations in the time domain*, SIAM J. Scientific Computing, 22 (2000), pp. 449–475. Cited on pages 37 and 82

[92] I. N. ONISHCHENKO, D. Y. SIDORENKO, AND G. V. SOTNIKOV, *Structure of electromagnetic field excited by an electron bunch in a semi-infinite dielectric-filled waveguide*, Phys. Rev. E, 65 (2002). Cited on page 133

[93] OPENMP. *http://www.openmp.org*. Cited on page 121

[94] S. PIPERNO, *Symplectic local time-stepping in non-dissipative dgtd methods applied to wave propagation problems*, tech. report, INRIA, No 5643, 2005. Cited on pages 89 and 146

[95] T. PLEWA, T. LINDE, AND V. WEIRS, eds., *Adaptive Mesh Refinement – Theory and Applications*, Springer, 2005. Cited on page 89

[96] M. B. PURSLEY, *Introduction to Digital Communications*, Pearson Prentice Hall, 2005. Cited on page 17

[97] W. REED AND T. HILL, *Triangular mesh methods for the neutron transport equation*, tech. report, Los Alamos Scientific Laboratory Report, 1973. Cited on page 89

[98] M. REISER, *Theory and Design of Charged Particle Beams*, Wiley & Sons, 1994. Cited on page 81

[99] J.-F. REMACLE, J. FLAHERTY, AND M. SHEPHARD, *An adaptive discontinuous galerkin technique with an orthogonal basis applied to compressible flow problems*, SIAM Review, 45 (2003), pp. 53–72. Cited on pages 89, 118, and 144

[100] J.-F. REMACLE, X. LI, N. CHEVAUGEON, AND M. SHEPHARD, *Transient mesh adaptation using conforming and non conforming mesh modifications*, in Proc. 11th International Meshing Roundtable, Sandia National Labs, 2002, pp. 261–272. Cited on pages 113 and 116

[101] R. F. REMIS, *On the stability of the finite-difference time-domain method*, Journal of Computational Physics, 163 (2000), pp. 249–261. Cited on page 71

[102] J. ROSSBACH AND P. SCHMÜSER, *Basic course on accelerator optics*, in CAS - CERN Accelerator School: 5th General accelerator physics course, S. Turner, ed., vol. I, 1992. Cited on page 16

[103] W. G. ROSSER, *Classical Electromagnetism Via Relativity*, Butterworth & Co, 1968. Cited on page 14

[104] C. RUNGE, *Über empirische funktionen und die interpolation zwischen äquidistanten ordinaten*, Zeitschrift für Mathematik und Physik, 46 (1901), pp. 224–243. Cited on page 93

[105] S. SCHNEPP, W. ACKERMANN, AND E. AREVALO, *Calculation of transverse and longitudinal wake potentials of thevalve and doublecross region*, PITZ Collaboration Meeting, DESY Zeuthen, June 12, (2007). Cited on page 141

[106] S. SCHNEPP, E. GJONAJ, AND T. WEILAND, *On the development of a self-consistent particle-in-cell (PIC) code using a time-adaptive mesh technique*, in Proc. 10th European Particle Accelerator Conference (EPAC), June 2006, pp. 2182–2184. Cited on pages 121 and 138

[107] ——, *A time-adaptive mesh approach for the self-consistent simulation of particle beams*, in Proceedings of the 9th International Computational Accelerator Physics Conference, Oct. 2006. Cited on page 139

[108] ——, *Analysis of a particle-in-cell code based on a time-adaptive mesh*, in Proceedings of the 2007 Particle Accelerator Conference, June 2007. Cited on pages 81, 138, and 139

[109] ——, *An hp-adaptive discontinuous galerkin method in time domain applied to the simulation of highly localized current sources*, in Annual Review of Progress in Applied Computational Electromagntics (ACES), 2008. Cited on pages 82 and 121

[110] S. SCHNEPP, S. SETZER, AND T. WEILAND, *Investigation of numerical noise in pic-codes*, in Proc. 9th Europ. Particle Acc. Conf. (EPAC), 2004. Cited on page 87

[111] P. SCHÜTT, *Zur Dynamik eines Elektronen-Hohlstrahls*, PhD thesis, Hamburg, 1988. Cited on page 82

[112] R. SCHUHMANN, F. MAYER, AND T. WEILAND, *Consistent 3d-FDTD subgrids for microwave applications*, in Proc. of the Int. Conf. on Electromagnetics in Adv. Appl., 2003. Cited on pages 87 and 146

[113] R. SCHUHMANN, P. SCHMIDT, AND T. WEILAND, *A new whitney-based material operator for the finite-integration technique on triangular grids*, IEEE Transactions on Magnetics (IEEE Mag), 38 (2002), pp. 409–412. Cited on page 97

[114] R. SCHUHMANN AND T. WEILAND, *Conservation of discrete energy and related laws in the finite integration technique*, Progress in Electromagnetic Research (PIER), 32 (2001), pp. 301–316. Cited on pages 35 and 95

[115] C. SCHWAB, *p- and hp-Finite Element Methods*, Oxford Science Publications, 1998. Cited on page 88

[116] S. SETZER, W. ACKERMANN, S. SCHNEPP, AND T. WEILAND, *Influence of beam tube obstacles on the emittance of the PITZ photoinjector*, in Proceedings of the 9th European Particle Accelerator Conference, July 2004, pp. 1981–1983. Cited on page 138

[117] P. SOLIN, T. VEJCHODSKY, AND M. ZITKA, *Modular hp-fem system hermes and its application to maxwell's equations*, Mathematics and Computers in Simulation, 76 (2007), pp. 223–228. Cited on pages 88 and 120

[118] E. SONNENDRÜCKER, F. FILBERT, E. FRIEDMAN, E. OUDET, AND J.-L. VAY, *Vlasov simulation of beams with a moving grid*, Comput. Phys. Comm., 164 (2004), pp. 390–395. Cited on page 78

[119] H. SPACHMANN, R. SCHUHMANN, AND T. WEILAND, *Higher order spatial operators for the finite integration theory*, in 18th Annual Review of Progress in Applied Computational Electromagnetics (ACES), vol. 17, 2002, pp. 11–22. Cited on page 47

[120] B. STEINER, W. ACKERMANN, S. FRANKE, W. MÜLLER, T. WEILAND, R. BARDAY, C. ECKHARDT, R. EICHHORN, J. ENDERS, C. HESSLER, Y. POLTORTSKA, A. RICHTER, AND M. ROTH, *Beam dynamics simulation of the new polarized electron injector of the s-dalinac*, in Proc. 11th Europ Particle Accelerator Conference (EPAC), 2008. Cited on page 78

[121] G. STRANG, *On the construction and comparison of difference schemes*, SIAM J. Numerical Analysis, 5 (1968), pp. 506–517. Cited on page 63

[122] S. SUWELACK, *Adaptivity strategies for the hp-adaptive discontinuous galerkin method*, master's thesis, TU Darmstadt, TEMF, 2008. Cited on pages 116, 117, 118, and 120

[123] B. SZABO AND I. BABUSKA, *Finite Element Analysis*, Wiley-Interscience, 1991. Cited on page 88

[124] A. TAFLOVE AND S. HAGNESS, *Computational electrodynamics: the finite-difference time-domain method*, Artech House, Boston, second ed., 2000. Cited on page 5

[125] TAMBCI, *A program for the self-consistent and low-dispersive simulation of arbitrarily shaped bunches in long accelerator structures*. Technical documentation: http://www.temf.de/downloads/tamBCI.pdf. Cited on page 121

[126] A. E. TAYLOR, *Introduction to Functional Analysis*, Wiley & Sons, 1958. Cited on page 48

[127] THE VISUALIZATION TOOLKIT (VTK). http://www.vtk.org. Cited on page 121

[128] P. THOMA, *Zur numerischen Lösung der Maxwellschen Gleichungen im Zeitbereich*, PhD thesis, TU Darmstadt, 1997. Cited on page 146

[129] P. THOMA AND T. WEILAND, *A consistent subgridding scheme for the finite difference time domain method*, Int. J. of Num. Modelling: Electronic Networks, Devices and Fields, 9 (1996), pp. 359–374. Cited on page 87

[130] C. ÜBERHUBER, *Computer Numerik*, vol. 1, Springer, 1995. Cited on page 93

[131] B. VAN LEER, *Towards the ultimate conservative scheme*, I-V (1973-1997). Cited on pages 37 and 94

[132] J.-L. VAY, P. COLELLA, A. FRIEDMAN, D. GROTE, P. MCCORQUODALE, AND D. SERAFINI, *Implementations of mesh refinement schemes for particle-in-cell plasma simulations*, Comput. Phys. Comm., 164 (2004), pp. 297–305. Cited on page 87

[133] J.-L. VAY, P. COLELLA, J. KWAN, P. MCCORQUODALE, D. SERAFINI, A. FRIEDMAN, D. GROTE, G. WESTENSKOW, J.-C. ADAM, A. HÉRON, AND I. HABER, *Application of adaptive mesh refinement to particle-in-cell simulations of plasmas and beams*, Physics of Plasma, 11 (2004), pp. 2928–2934. Cited on page 87

[134] J. VILLASENOR AND O. BUNEMAN, *Rigorous charge conservation for local electromagnetic field solvers*, Comput. Phys. Comm., 69 (1992), pp. 306–316. Cited on pages 145 and 149

[135] R. WANZENBERG AND T. WEILAND, *Wakefields and impedances*, in Proc. CAT-CERN Accelerator School (CCAS), vol. 1, 1993, pp. 140–180. Cited on page 79

[136] T. WEILAND, *Eine methode zur lösung der maxwellschen gleichungen für sechskomponentige felder auf diskreter basis*, Electronics and Communication (AEÜ), 31(3) (1977), pp. 116–120. Cited on pages 5 and 29

[137] ———, *On the numerical solution of maxwell's equations and applications in the field of accelerator physics*, Particle Accelerators, 15 (1984), pp. 245–292. Cited on pages 29 and 31

[138] ———, *On the unique numerical solution of maxwellian eigenvalue problems in three dimensions*, Particle Accelerators, 17 (1985), pp. 227–242. Cited on pages 33 and 52

[139] ———, *Time domain electromagnetic field computation with finite difference methods*, Int. J. of Num. Modelling: Electronic Networks, Devices and Fields, 9 (1996), pp. 295–319. Cited on pages 33 and 58

[140] H. WIEDEMANN, *Particle Accelerator Physics - Basic Principles and Linear Beam Dynamics*, vol. I, Springer, 1999. Cited on page 19

[141] ———, *Particle Accelerator Physics - Nonlinear and Higher-Order Beam Dynamics*, vol. II, Springer, 1999. Cited on pages 16 and 21

[142] K. WILLE, *The Physics of Particle Accelerators*, Oxford University Press, 2000. Cited on pages 16 and 18

[143] F. WOLFHEIMER, E. GJONAJ, AND T. WEILAND, *Parallel particle-in-cell (pic) codes*, in Proc. of the 9th Int. Comp. Acc. Phys. Conf. (ICAP), 2006. Cited on page 138

[144] K. YEE, *Numerical solution of initial boundary value problem involving maxwell's equations in isotropic media*, IEEE Trans. on Antenna and Prop., 14 (1966), pp. 302–307. Cited on pages 23 and 30

[145] H. YOSHIDA, *Construction of higher order symplectic integrators*, Phys. Rev. Letters A, 150 (1990), pp. 262–268. Cited on page 58

[146] ———, *Recent progress in the theory and application of symplectic integrators*, Celestial Mechanics and Dynamical Astronomy, 56 (1993), pp. 27–43. Cited on pages 35, 58, and 63

[147] I. ZAGORODNOV, R. SCHUHMANN, AND T. WEILAND, *A uniformly stable conformal FDTD-method in cartesian grids*, International Journal of Numerical Modelling: Electronic Networks, Devices and Fields (NumMod), Vol. 16 (2003), pp. 127–141. Cited on page 27

[148] ———, *A simplified conformal (SC) method for modeling curved boundaries in fdtd without time step reduction*, in Proceedings of the IEEE MTT-S International Microwave Symposium (IMS), 2006, pp. 177–180. Cited on pages 27 and 36

[149] M. ZHANG AND P. SCHUETT, *TESLA FEL gun simulations with PARMELA and MAFIA*, in Proc. Comp. Acc. Phys. Conf. (CAP), 1996. Cited on pages 138 and 139

[150] O. C. ZIENKIEWICZ AND Y. K. CHEUNG, *The Finite element method in structural and continuum mechanics*, McGraw-Hill, 1968. Cited on pages 5 and 33

Die VDM Verlagsservicegesellschaft sucht für wissenschaftliche Verlage abgeschlossene und herausragende

Dissertationen, Habilitationen, Diplomarbeiten, Master Theses, Magisterarbeiten usw.

für die kostenlose Publikation als Fachbuch.

Sie verfügen über eine Arbeit, die hohen inhaltlichen und formalen Ansprüchen genügt, und haben Interesse an einer honorarvergüteten Publikation?

Dann senden Sie bitte erste Informationen über sich und Ihre Arbeit per Email an *info@vdm-vsg.de*.

Sie erhalten kurzfristig unser Feedback!

VDM Verlagsservicegesellschaft mbH
Dudweiler Landstr. 99
D - 66123 Saarbrücken
Telefon +49 681 3720 174
Fax +49 681 3720 1749
www.vdm-vsg.de

Die VDM Verlagsservicegesellschaft mbH vertritt

Printed by Books on Demand GmbH, Norderstedt / Germany